NEW PATHWAYS IN HIGH-ENERGY PHYSICS II

New Particles – Theories and Experiments

Studies in the Natural Sciences

A Series from the Center for Theoretical Studies

University of Miami, Coral Gables, Florida

A Continuation Order Plan is available for this series. A continuation order will bring delivery of each new volume immediately upon publication. Volumes are billed only upon actual shipment. For further information please contact the publisher.

NEW PATHWAYS IN HIGH-ENERGY PHYSICS II

New Particles — Theories and Experiments

Edited by

Arnold Perlmutter

Center for Theoretical Studies
University of Miami
Coral Gables, Florida

PLENUM PRESS ● **NEW YORK AND LONDON**

Library of Congress Cataloging in Publication Data

Orbis Scientiae, University of Miami, 1976.
  New pathways in high-energy physics.

  (Studies in the natural sciences; v. 10-11)
  Includes indexes.
  1. Particles (Nuclear physics)—Congresses. I. Perlmutter, Arnold, 1928-   II.
Miami, University of, Coral Gables, Fla. Center for Theoretical Studies. III. Title.
IV. Series.
QC793.O7 1976   539.7'6   76-20476
ISBN 0-306-36911-7 (v. 2)

A part of the Proceedings of Orbis Scientiae 1976 held by the
Center for Theoretical Studies, University of Miami, Coral Gables, Florida,
January 19-22, 1976

© 1976 Plenum Press, New York
A Division of Plenum Publishing Corporation
227 West 17th Street, New York, N.Y. 10011

Printed in the United States of America

Preface

This year, Orbis Scientiae 1976, dedicated to the Bicentennial of the United States of America, was devoted entirely to recent developments in high energy physics. These proceedings contain nearly all of the papers presented at Orbis, held at the Center for Theoretical Studies, University of Miami, during January 19-22, 1976.

The organization of Orbis this year was due mainly to the moderators of the sessions, principally Sydney Meshkov, Murray Gell-Mann, Yoichiro Nambu, Glennys Farrar, Fred Zachariasen and Behram Kursunoglu, who was also chairman of the conference. The coherence of the various sessions is due to their efforts, and special thanks are due to Sydney Meshkov who was responsible for coordinating many of the efforts of the moderators and for including essentially all of the frontier developments in high energy physics.

Because of the number of papers and their integrated length, it has been necessary to divide these proceedings into two volumes. An effort has been made to divide the material in the two volumes into fundamental questions (including the appearance of magnetic charge in particle physics) and recent high energy results and attendant phenomenology.

These volumes were prepared by Mrs. Helga Billings, Mrs. Elva Brady and Ms. Yvonne Leber, and their dedication and skill are gratefully acknowledged. Their efforts

during Orbis were supplemented by those of Mrs.
Jacquelyn Zagursky, with our appreciation.   The photo-
graphs were taken by Ms. Shirley Busch.

Orbis Scientiae 1976 received some support from
the National Science Foundation Office of International
Programs and Energy Research and Development Administration.

The Editors

Contents of Volume 11

Contents of Volume 10

Participants of the Orbis Scientiae 1976, January 19-22, 1976

New Pathways in High Energy Physics

NEW PARTICLE SPECTROSCOPY AND DECAYS[*]

Frederick J. Gilman

Stanford Linear Accelerator Center

Stanford University

Stanford, California 94305

I. INTRODUCTION

In the year since the last meeting in this series, great progress has been made in the spectroscopy of the new particles and their decays. Much of this progress is either directly the result of experiment or at least has been very much stimulated by the astonishing results presented to us one after another by our experimental colleagues. In one way, what has happened is exemplified by the contrast between what was known about the 4-GeV region in $e^+e^- \rightarrow$ hadrons a year ago[1] (Figure 1) and data[2] which were shown this morning (Figure 2).

To my mind, the most shattering of the developments in the past year are the events[3] of the form, $e^+e^- \rightarrow e^{\pm} + \mu^{\mp} + (\geq 2$ unobserved particles), inasmuch as they are consistent with, or even point toward, a pair produced

---

[*]Work supported by the U. S. Energy Research and Development Administration.

1

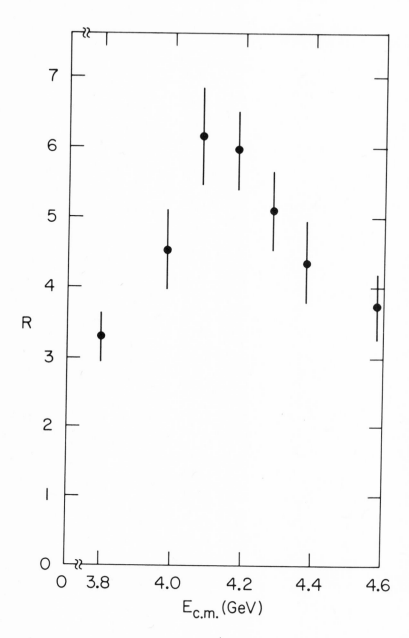

Figure 1: Values of $R \equiv \sigma(e^+e^- \to hadrons)/\sigma(e^+e^- \to \mu^+\mu^-)$ near $E_{cm} = 4$ GeV as of one year ago.[1]

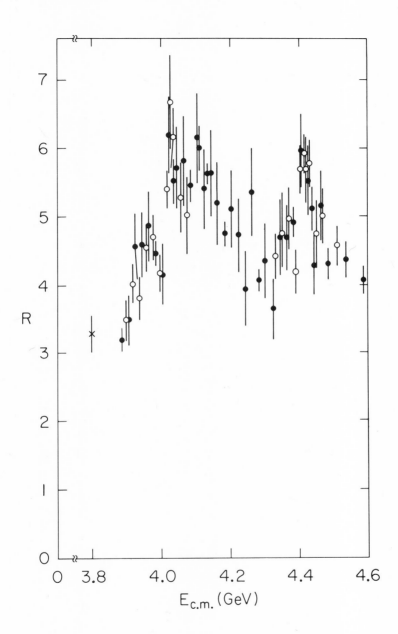

Figure 2: Values of R near E_{cm} = 4 GeV presented[2] to this conference.

charged heavy lepton as their origin. However, as there
is no dramatic movement either experimentally or theoretically on this subject in the past few months, I will lay
this topic aside for the remainder of this talk with only
the remark that no conventional explanation of these
events has been found and everything remains consistent
with their being due to a plain, ordinary, garden-variety
heavy lepton!

Instead, I should like to discuss where we stand
phenomenologically on the spectroscopy and decays of
new hadrons, both those possibly carrying a new quantum number ("charmed particles") and those without it.
Such a discussion is important for a number of reasons.
First, it teaches us about the nature of the new particles
themselves: are they hadrons and exhibit a hadronic
spectroscopy, do some of them carry a new quantum number,
is there more than one new quark, etc? Second, we learn
about the existence of other new states by studying the
decay modes of the known states, as has happened already
with the states below the ψ' reached from it by gamma
ray emission. Also, although I will not discuss it today,
there are hints that there may be things to be learned
even about the "old" spectroscopy by studying decays of
the new particles. Third, we can study the transformation
properties of the new states under isospin or SU(3) if
these symmetries are preserved in their decays and thereby
transmitted to the final state hadrons. Finally, there
is much to learn about dynamics, ranging from the "Zweig
rule" and the calculation of masses and transition amplitudes in the (non-relativistic?) quark model, to
Adler zeros and the usefulness of vector dominance.

In a number of places I will emphasize difficulties
or problems in our understanding as well as gaps in the
information available to us. Indeed, in looking back
at the talk I gave last year[4] it turns out rather surprisingly to be the case that several of the most important problem areas are the same, which troubles me
somewhat. So if this afternoon's session is in "psychotherapy," I'm afraid the audience will have to play
doctor while I'm the patient. The emphasis on extant
problems is simply because by understanding their solution we will all make considerable progress. That way,
we can hope that the doctor-patient roles can be largely
reversed at the next conference.

II. THE SPECTROSCOPY OF THE ψ's AND THEIR RELATIVES

It is very useful to have a model of the new particles
in the back of one's mind as a reference point when discussing their spectroscopy and decays. For such a model
we take the hypothesis that in addition to the u, d,
and s quarks and corresponding antiquarks, which are
supposed to be the basic constituents of hadrons, there
is one (or more) new quark, c. This quark(s) is assumed
to carry a new quantum number(s), with the generic name
charm,[5] which is to be conserved in strong and electromagnetic interactions. Until the last section of this
talk wherein charmed particles and their decays are
discussed, the specific charm quantum number of Glashow
et al.[6] will have no distinctive role to play vis-à-vis
any other new quantum number conserved in strong and
electromagnetic processes.

The ψ, ψ',... are taken to be $c\bar{c}$ bound states or
resonances. Such states with a mass less than about
4 GeV cannot decay into a pair of charmed particles for

kinematic reasons, and hence only have "ordinary" mesons
and baryons among their hadronic decay products. Since
the quarks in the $c\bar{c}$ state then do not appear within the
final hadrons, such decays are forbidden by the "Zweig
rule"[7] and the corresponding widths are very much suppressed (by a factor $\sim 10^4$) from those of an ordinary
hadron with such a mass.

Spectroscopically, one expects a set of SU(3)
singlet (and therefore SU(2) singlet) $c\bar{c}$ states,
given that the charmed quark is itself an SU(3) singlet.
The lowest mass such states would have zero orbital
angular momentum (L) between the quark and antiquark,
and hence be $J^{PC}=0^{-+}$ and 1^{--} states. The L=1 states,
$0^{++}, 1^{++}, 2^{++}$ and 1^{+-} should lie several hundred MeV
higher. Following this would be the L=2 states (1^{--},
$2^{--}, 3^{--}$ and 2^{-+}) and/or radially excited L=0 states.

For comparison, the presently known mass, X(2.8)
has been reinforced by the new data[8] presented here
from DESY.[9] Although all we know is that $J \neq 1$ and
C = +, it is conventional in the $c\bar{c}$ scheme to assign
this state $J^{PC}=0^{-+}$ quantum numbers so that it is the
quark spin S=0 ground state partner of the $\psi(3.095)$
$(\equiv\psi)$ with S=1.

Between the ψ and ψ' ($\equiv \psi(3.684)$) are the C=+ states,[9]
$\chi(3.41)$, $P_c(3.51)$ and $\chi(3.53)$. All are found in decays
of the ψ' involving emission of a gamma ray. The $\chi(3.41)$
and $P_c(3.51)$ have widths consistent with experimental
resolution; moreover, the dominant electromagnetic decay
of P_c into $\gamma\psi$ points to a narrow width. On the other
hand, $\chi(3.53)$ is observed (in hadronic decay modes) to
be wider than resolution and to have a central mass value
different from the P_c. One then infers the $\chi(3.53)$ is
more than one state (if they are narrow) and economy in

the proliferation of states then suggests that the P_c be identified with one of these states at the lower end of the mass range subsumed in the $\chi(3.53)$. The other(s) must lie at somewhat higher mass to give the impression of a broader state. There are then at least three C=+ states between the ψ and ψ'.

Above the ψ' (3.684) is the broader structure "$\psi(4.1)$" and the $\psi(4.414)$, discussed[2] this morning. Considering the leptonic width of the $\psi(4.414)$, which is proportional to the area under the resonance bump in a plot of $\sigma(e^+e^- \rightarrow$ hadrons) vs E_{cm}, and the sensitivity of previous scans for resonances, one concludes that objects like the $\psi(4.414)$ could exist at almost any mass and have escaped detection up to now. The apparent structure within the "$\psi(4.1)$", particularly the jump in the cross section by ~ 50% over ~ 20 MeV in E_{cm} near 4 GeV, strongly suggests that several objects like the $\psi(4.414)$ are to be found in this region. We probably have entered a new regime of "mini-structure"- i.e., bumps whose area is one-twentieth or less than that of the ψ. In particular, further resonances with leptonic widths comparable to that of $\psi(4.414)$, although likely with larger total widths, seem a foregone conclusion above ~ 4.6 GeV.

The present spectrum of states is <u>consistent</u> with the spectrum expected from <u>one</u> new quark bound to its antiquark. In particular, the X(2.8) and $\psi(3.095)$ are the L=0 states, while the χ's and P_c are good candidates for the C=+, L=1 states as indicated in Fig. 3.

However, at a minimum, the dynamics of the $c\bar{c}$ system must be complicated to understand the "mini-structure" in the 4- GeV region. Simple non-relativistic potentials would seem inadequate, given the number of states which are very likely present there. A number of proposals[10]

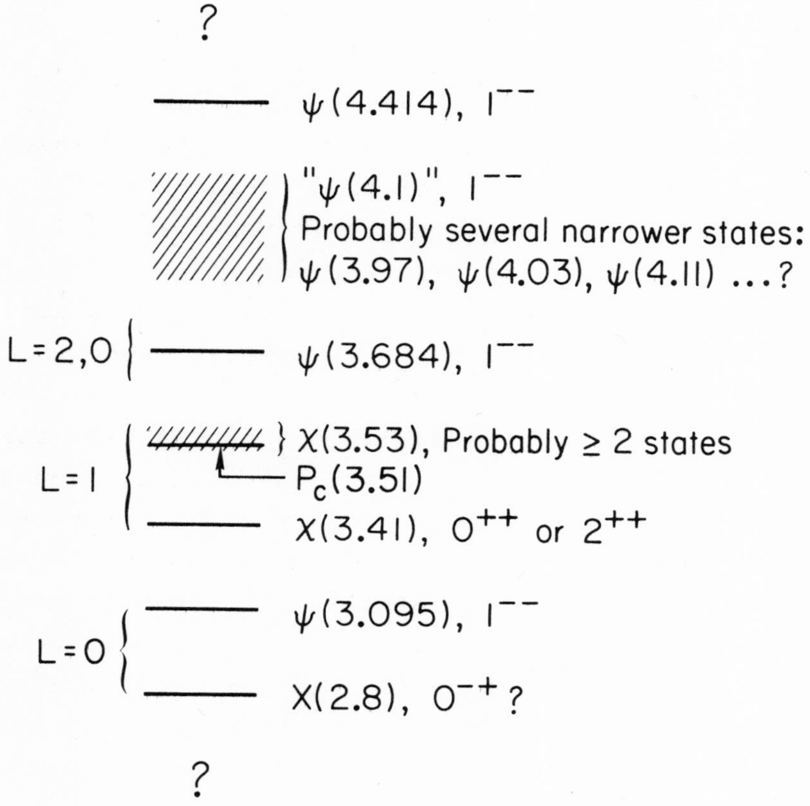

?

——— $\psi(4.414)$, 1^{--}

//////// $"\psi(4.1)"$, 1^{--}
//////// Probably several narrower states:
//////// $\psi(3.97)$, $\psi(4.03)$, $\psi(4.11)$ ...?

$L = 2, 0$ ——— $\psi(3.684)$, 1^{--}

$L = 1$ ///////// } $X(3.53)$, Probably ≥ 2 states
——— $P_c(3.51)$
——— $X(3.41)$, 0^{++} or 2^{++}

$L = 0$ ——— $\psi(3.095)$, 1^{--}
——— $X(2.8)$, 0^{-+}?

?

Figure 3: Known spectroscopy of the ψ's and related
           states.

to explain this situation by involving more complicated
quark configurations have already been made. Alternately,
one may invoke the existence of another new quark, somewhat heavier than the c quark, and with its bound states
with the corresponding antiquark having a small admixture
of $c\bar{c}$ so as to permit decay into pairs of charmed
particles with an almost "normal" hadronic width.

Aside from the 4-GeV region, it is possible that
a second new quark exists and its corresponding spectroscopy is accessible to present experiments. In particular,
if this quark had charge - 1/3 while that bound in the
ψ's has charge +2/3, the corresponding lowest mass vector
meson might have gone undetected[11] in the SPEAR scan[12]
for narrow resonances provided its mass was above ~ 5 GeV.
The fragility of the present situation with respect to
consistency with what is expected from only one new quark
is to be noted in general. The existence of other narrow
states below X(2.8), the existence of other than just the
specific L = 1 states and the pseudoscalar partner of the
ψ' between 3.1 and 3.7 GeV, or the existence of further
very narrow states above ~ 4 GeV would immediately call
for the introduction of more new quarks or, depending
on the character of the hypothetical additional states,
even the possible abandonment of the whole picture.

III. $\psi(3.095)$ DECAYS

The $\psi(3.095)$ has $J^{PC}=1^{--}$ and has both decays through
one photon and "direct" decays. Decays proceeding through
one photon[13] into e^+e^- and $\mu^+\mu^-$ each comprise ~ 7% of the
total width, and imply the existence of $\psi \to \gamma_v \to$ hadrons with
a branching ratio of $R_{off-resonance} \times$ 7% $\simeq$ 17%. From
study of a number of exclusive decay channels, there is

strong evidence that the ψ acts as G=−, I=0 object in its
direct decays.[14]

The totaling-up of all the observed or inferred decays of the ψ involving hadrons (plus possible gamma rays)
has changed little since the summer.[15] The arithmetic
goes as follows (in percent of the ψ decays involving
hadrons):

$$
\begin{array}{llll}
\psi & \rightarrow & 3\pi,\ 5\pi,\ 7\pi,\ldots & 20\text{-}30\% \\
 & \rightarrow & K\bar{K} + \pi\text{'s} & 20\text{-}30\% \\
 & \rightarrow & N\bar{N} + \pi\text{'s} & 5\text{-}10\% \\
 & \rightarrow & \gamma_v \rightarrow \text{hadrons} & 20\% \\
 & \rightarrow & \gamma + X(2.8) & 2\text{-}10\% \\
 & & & \overline{\qquad\qquad} \\
 & & & 67\text{-}100\%
\end{array}
$$

$\left.\begin{array}{l}20\text{-}30\% \\ 20\text{-}30\% \\ 5\text{-}10\%\end{array}\right\}$ direct decays

Still to be included are some modes containing γ's,η's,
etc. The upper limit on the photon decay into the X(2.8)
is a relatively conservative one based on the absence of
monochromatic gamma rays,[16] while the lower number assumes
that the decay X $\rightarrow p\bar{p}$ is real with such a mode being at
most 1% of all X $\rightarrow$ hadron decays in any reasonable model.
In any case, what is to be learned from this exercise is
not that we understand where 100.00% of ψ decays go: One
cannot rule out another 10 or 20% mode or modes involving
multineutrals a large part of the time (e.g., $\eta'\omega$).
Rather, one learns that a fairly healthy fraction is
accounted for as rather inauspicious direct decays into
hadrons and that major (∼40%, corresponding to $\Gamma \sim 25$ keV)
unconventional modes are not possible.

Another area where little has changed since this past
summer is the question of the SU(3) character of the ψ.
If composed of SU(3) singlet quarks, the ψ should be a

singlet. If SU(3) is conserved in the direct decay process, then an examination of relative decay rates into
specific channels will reflect on the character of the
ψ itself. In particular,[17] $\psi \rightarrow$ KK* and K*K** are observed,[18] while decays into $K_S K_L$ (or K^+K^-), K*K*, K**K**,
and KK** are not, which is just the way an SU(3) singlet
state should behave. In other words, where there are
zeros in the SU(3) Clebsch table for decays of an SU(3)
singlet into two mesons, one finds no evidence for such
channels in ψ decays. A different test involves the
ratio of rates for two allowed processes. Here the one
measured example is $\Gamma(\psi \rightarrow \pi^+\rho^-)/\Gamma(\psi \rightarrow K^+K^{*-})$, which is
found to be ~ $2\frac{1}{2}$ rather than unity as expected for a
singlet. Note that this failure is <u>not</u> attributable to
contamination[19] of the direct decays by $\psi \rightarrow \gamma_v \rightarrow \pi^+\rho^-$
and $\psi \rightarrow \gamma_v \rightarrow K^+K^{*-}$, for the ratio is still unity when these
processes are included as well, if SU(3) holds for the
relevant photon-hadron vertex. At the moment then, the
situation is confused. It is possible that we will have
to face ~ 50% violations of SU(3) in the amplitudes to
various channels-presumably induced by SU(3) violation
in the decay process if we wish to continue to believe
the ψ is an SU(3) singlet. However, with the recent
tripling of the data more accurate versions of previous
tests as well as new tests in other channels will become
possible. Perhaps we should wait for these results on
both the ψ and ψ', before coming to a definite conclusion
about SU(3) for the new particles and their decays.

IV. $\psi(3.684)$ DECAYS

A major development in decays of the new particles
over the past year has been the discovery of the gamma ray
decays of the ψ' into C=+ intermediate states.[9] A

relatively minor consequence of this is that the "anything" in $\psi' \to \psi$ + anything is now completely consistent
with being accounted for[20] by $\pi^+\pi^-$, $\pi^0\pi^0$, η and $\gamma\gamma$.
The $\pi \pi \psi$ and $\eta\psi$ modes demand that the ψ and ψ' have
the same isospin and G parity. Aside from being squeezed
out by the known modes, other specific channels like
$\psi' \to \pi^0 \psi$, which are allowed .in some models, now have
very stringent upper limits placed upon them.[20]

The more accurate measurement of the branching ratio
for $\psi' \to \eta \psi$ of 4.3 ± 0.8% now available[20] permits one to
quantitatively check another aspect of the dynamics, that
of vector dominance involving the ψ and ψ'. For example,
ψ-dominance of the photon in $\psi' \to \eta \gamma$ leads one to expect[2?]

$$\Gamma(\psi' \to \eta\gamma)/\Gamma(\psi' \to \eta\psi) \simeq 0.5.$$

However, employing the upper bound from DESY[9] on $\psi' \to \eta\gamma$,
one finds

$$\Gamma(\psi' \to \eta\gamma)/\Gamma(\psi' \to \eta\psi) \leq 0.14\%/4.3\% \approx 0.03,$$

so that the theoretical prediction is too large by over
an order of magnitude. In a completely analogous way,
ψ' dominance of the photon in $\psi \to \eta\gamma$ leads to a predicted[2?]
width for this process of roughly 1 keV. This is too
large by an order of magnitude: experiment gives a value[9]
of ~ 0.1 keV.

Of course, one is extrapolating a very long way from
the photon to the ψ and ψ' mass shell, and there are other
heavy vector mesons[22] which contribute to each amplitude
which haven't been taken into account. But this is precisely the point: the failure of these most naive calculations should be taken as a warning against relying on

the same exercise done on other amplitudes. A particular case in point is the extraction of $\sigma_T(\psi N)$ from $d\sigma/dt(\gamma N \rightarrow \psi N)$ by assuming the photoproduction amplitude is mostly imaginary (diffractive) and then using ψ-dominance of the photon. There is no a priori reason for vector dominance to work much better here than in the two cases discussed above. Fortunately, a measurement of $\sigma_T(\psi N)$ independent of any vector dominance assumption is possible by studying the A dependence of the cross section on nuclei. This is now in progress.[23]

Some new developments have occurred with respect to "direct" decays of the ψ' into ordinary hadrons. Enough such decays have been seen[18] so that a pattern is beginning to emerge with respect to the same decays of the ψ: it is that $\Gamma(\psi' \rightarrow$ hadronic channel$) = (\frac{1}{2}$ to $\frac{1}{5}) \times \Gamma(\psi \rightarrow$ hadronic channel$)$ for each of the "direct decay" channels so far found. If the pattern is general then we would have

$$\Gamma(\psi' \rightarrow \text{hadrons})\Big|_{\substack{\text{direct} \\ \text{decays}}} = (\tfrac{1}{2}\text{to}\tfrac{1}{5})\Gamma(\psi \rightarrow \text{hadrons})\Big|_{\substack{\text{direct} \\ \text{decays}}} .$$

This is of some importance for it checks against another pair of measured widths

$$\Gamma(\psi' \rightarrow e^+ e^-) \simeq \frac{1}{2.3}\Gamma(\psi \rightarrow e^+ e^-).$$

In the charmonium picture both the $e^+ e^-$ decay and the "direct" decay are proportional to the square of the wave function of the state at the origin, $|f(0)|^2$. The consistency of the two independent measurements[24] of the ratio of the square of the ψ and ψ' wave functions at the origin provides some encouragement to this picture of "direct" and e^+e^- decays.

We are now in a position to add up the known or inferred decay modes of the ψ' involving hadrons:

$$
\begin{aligned}
\psi' &\to \psi + \text{anything} & \sim 57\% \\
&\to \gamma_V \to \text{hadrons} & 3\% \\
&\to \text{"direct" hadrons} & \lesssim 10\% \\
&\to \gamma + \chi \\
&\qquad \hookrightarrow \text{hadrons} & 5\text{-}10\% \\
\hline
& & 75\text{-}80\%
\end{aligned}
$$

The 10% number for direct decays comes from taking all ψ decays other than those proceeding through one photon and scaling them by the ratio of the square of the ψ' and ψ wave functions at the origin, as measured in their e^+e^- decays. It is therefore presumably an upper limit. The estimate of 5-10% for gamma ray decays ending in known χ states decaying into hadrons is based on scaling up the observed $\chi \to$ hadron modes to guess their total direct hadronic decays.

The remaining 20-25% of unaccounted for ψ' modes containing hadrons is a serious discrepancy, unlike the case for the superficially similar analysis for the ψ. For in this case it is not "ordinary" direct decays which have not been explicitly reconstructed which might fill the void: such direct decays are already rather liberally accounted for in the 10% figure obtained by scaling down all possible "direct" ψ decays to the ψ' by the ratio of their leptonic widths.

The discrepancy can also not be entirely due to the decay $\psi' \to \omega + X(2.8)$, as we heard this morning.[2] Furthermore, if we assume the $\chi(3.41)$ is the 0^{++} p-wave state

with the others lying above 3.50 GeV, we can use the
upper bound[25] on $\psi' \to \gamma + \chi(3.41)$ to bound the size of the
remaining transitions to the 1^{++} and 2^{++} states. Even
taken altogether they cannot fill up the gap of unaccounted for ψ' decays. Similarly, the bounds[25] on any single
monochromatic photon transition prevent the decay of the
ψ' into its pseudoscalar partner (if not already seen as
a χ or P_c state) by gamma ray emission from accounting for
the total discrepancy.

However, it is still in the range of possibility that
the problem will be solved by each of several (of the
above?) channels eating up several percent of the ψ'
decays, leaving any remaining discrepancy within the
statistical errors on the data. Another, relatively
conventional, possibility is that some important modes
which exist for both the ψ and ψ' do not scale as we have
done for all "direct decays." An explicit example is
provided by assuming that the η and/or η' have a small
$c\bar{c}$ component, as has been proposed by several authors.[26]
The new round of ψ and ψ' decay experiments at SPEAR may
give us a clue as to the direction in which the answer
lies.

V. $\chi, P_c, X, \ldots$ DECAYS

In their hadronic decays as so far observed, the χ's
behave as would be expected for $C=G=+$ states: formed by
emission of one photon from the ψ', decay channels with
even numbers of pions are observed. Some particular decay
modes which are already accessible are of special importance for the determination of quantum numbers. Decay of
a $C=+$ object into two pseudo-scalars implies J is even and
therefore parity $P=+$. If both $\pi^+\pi^-$ and K^+K^- are present,
as indicated[9] for the (3.41), then $I=0$, for the $\pi^+\pi^-$ sy-

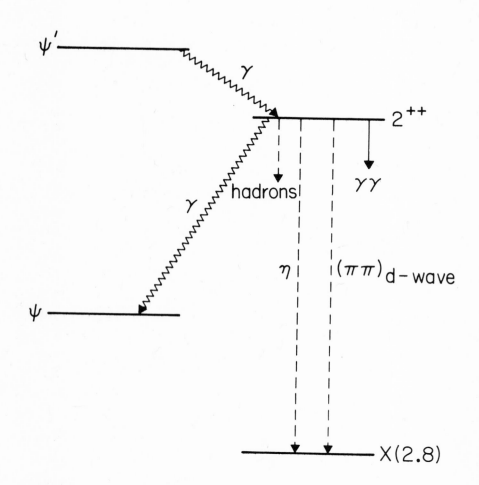

Figure 4: Formation and some possible decays of a $c\bar{c}$, $J^{PC} = 2^{++}$ state.

stem may have I=0 or 2 while K^+K^- has I=0 or 1. The
assignment of the $\chi(3.41)$ to $I^G=0^+$ and $J^{PC}=(even)^{++}$ is
probably the most important observation on the intermediate
states between the ψ and ψ' up to this point, inasmuch as
it both rules out a pseudoscalar state and is just what is
expected for the L=1, 0^{++} or 2^{++} $c\bar{c}$ states.

Observation of decays like πA_2, $K\bar{K}\pi$, etc. are
likely to be of use in the future since they immediately
rule out the assignment $J^P=0^+$. And of course, $\gamma\gamma$, as
observed[9] for the X(2.8), rules out J=1.

Below the ψ', the three L=1 states $0^{++},1^{++},2^{++}$
should be found arising from monochromatic gamma ray
decays of the ψ'. The 0^{++} state lies lowest in most
models and so is assumed to be the $\chi(3.41)$. The formation, and some possible decays of the 2^{++} state, which
might be contained in the $\chi(3.53)$, are shown in Fig. 4.
The decay of such a state into X(2.8) could well be
competitive with the direct decay into ordinary hadrons.

A fourth C=+ state should be found between ψ and ψ':
the pseudoscalar partner of the ψ'. Some possibilities
for formation and decay of the pseudoscalar states are
shown in Fig. 5. Again the transition from the heavier
to lighter pseudoscalar may be non-negligible in comparison to other decay modes of the upper 0^{-+} state. In
the charmonium model direct decays into hadrons are supposed to have widths in the multi-MeV range[27] and hence
would likely dominate the remaining modes shown in Fig. 5.

Above the ψ' it is possible that there are further
very narrow states until one gets to the threshold for
decay into a pair of charmed hadrons. This might be as
high as 3.9 GeV, in which case other L=2 or even L=3
states of the $c\bar{c}$ system would have widths like the ψ
and ψ'.

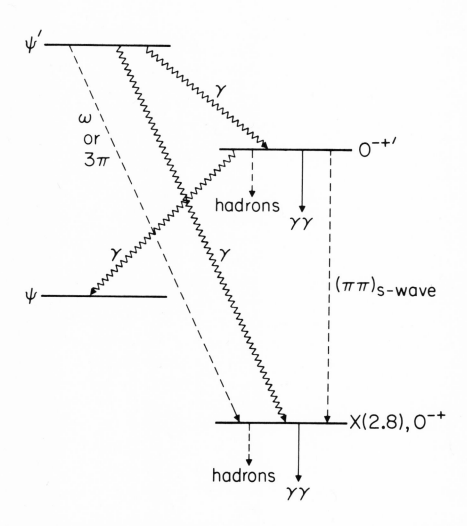

Figure 5:   Some possibilities for formation and decay of
            pseudoscalar partners of the ψ and ψ'.

Unfortunately, such states will be very difficult to detect experimentally, aside perhaps from the 1^{--} state with L=2 which could couple to e^+e^- with enough strength to be seen as a bump in the total cross section. One possible way to try to find some of these states is by looking for

$$e^+e^- \to \psi + (C=+)$$

at the center-of-mass energies above ~ 7 GeV. The ψ is readily detectable in the e^+e^- or $\mu^+\mu^-$ mode, but the absence[28] of an inclusive ψ signal down to a level of ~ 1% of the total cross section means that such processes are quite rare, at best.

The C=-1 states might be accessible by studying

$$\psi(4.?) \to (\pi\pi \text{ or } \eta) + (C=-1),$$

where $\psi(4.?)$ is one of the bumps in the 4 GeV region. Since several of these bumps have apparent widths of 20-40 MeV, and since $\psi' \to \pi\pi\psi$ has a partial width of ~100 keV, it might be possible that such decays occur at the 1% branching ratio level for a given $\psi(4.?)$. This method is also applicable to finding the quark spin singlet, p-wave state with $J^{PC}=1^{+-}$ which presumably lies between the ψ and ψ', near the other L=1 states. However, it cannot be formed by gamma emission from the ψ' because of charge conjugation invariance, while phase space presumably stops the formation by emission of $\pi\pi$ from the ψ'. If some of the other L=1 states with C=+ lie above it they can decay into it by emitting a gamma ray as indicated in Fig. 6. As also shown there, such a state could have a number of interesting competitive modes of

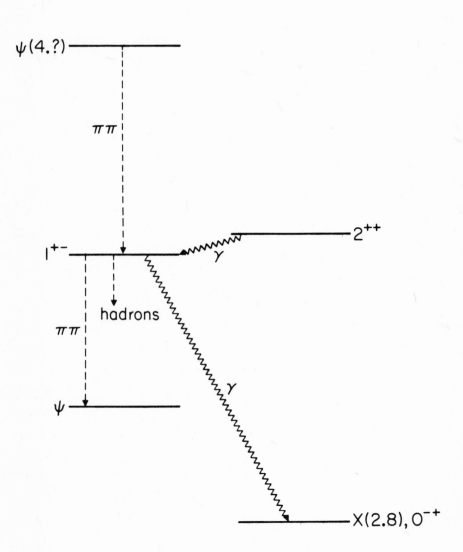

Figure 6:   Some possibilities for formation and decay of
            a $J^{PC} = 1^{+-} c\bar{c}$ state between ψ and ψ'.

decay.

VI. CHARMED PARTICLES

Particles containing only one new quark (antiquark) carry 1(-1) unit of charm. The lowest mass such particle (meson?) should lie between 1.84 GeV and 1.95 GeV. The lower limit arises from the narrow width of the ψ', while the upper limit is based on the rapid rises and falls in R starting at 3.9 GeV, presumably due to non-narrow $c\bar{c}$ resonances decaying into pairs of charmed particles with ordinary hadronic widths.

Further evidence of the existence of hadrons carrying a new quantum number comes from the dimuon events induced by neutrinos[29,30] together with the Gargamelle[31] and Fermilab[32] bubble chamber events of the form

$$\nu_\mu N \rightarrow \mu^- + e^+ + (Vee)^0 + \ldots \qquad .$$

It is difficult to find any explanation for such events other than that a new heavy hadron is being produced which decays weakly (i.e. semi-leptonically), but "promptly" enough so that the position appears to originate at the interaction vertex. With a non-negligible branching ratio for weak decays, the new hadrons must be forbidden from decaying strongly or electromagnetically by posessing a quantum number conserved in these interactions. If the semileptonic branching ratio of such a particle is $\sim$ 10%, the present data on, e.g., $e^+ e^- \rightarrow \mu + K_S + \ldots$, are not yet sensitive enough[28] to see such a signal for charm production unambiguously, even if final states containing charmed particles are a fair fraction of all events above ~ 4 GeV in E_{cm}.

The critical place to look for evidence of charm production in e^+e^- annihilation is the 4 GeV region. Indeed, much more important than exactly how many states like the $\psi(4.414)$ exist, is the use they can be put to in delineating the properties of charmed particles. For if such resonances are $c\bar{c}$ states with widths of 20-40 MeV because they are decaying into pairs of charmed particles (+ other hadrons), then any change in $\langle n_{ch}\rangle$, $\langle K_S\rangle$, $\langle\gamma\rangle$, $\langle\mu\rangle$, etc., etc., off and on a bump in R is assignable to the effects due to (pairwise) charm production and subsequent weak decay.

Since, e.g., R changes by ~ 50% in 20 MeV in E_{cm} near 4.03 GeV, such an analysis is <u>independent</u> of the existence <u>of any other new</u> (or old) <u>physics which varies slowly with energy</u>, such as the existence of pair produced heavy leptons in the same energy region. Note that (Fig. 2) <u>off the bumps</u> in the 4 GeV region R≃4, so that if a charged heavy lepton does exist with M_L≈1.8 GeV, <u>there is less than about one unit of R available for charmed particle production</u> after taking away the "old physics" (R~1) contributions.

In some ways we have come almost full circle since the conference[4] one year ago with regard to searches for charmed particles in e^+e^- annihilation. At the time of the last conference in this series searches[33] for bumps in invariant mass plots of two and three body systems produced in e^+e^- annihilation at E_{cm}=4.8 GeV showed no statistically significant evidence for charmed particles decaying non-leptonically. One way out of this was to assume that the lowest mass charmed particle typically decayed into relatively high multiplicity states. However, this could not be, for the observed charged multiplicity on entering the 4 GeV region, where R approximately doubled, showed no great jump and was ~4. Since one had two charmed particles in every "new physics" event, this created a "multiplicity crunch."

By mid-summer the crunch was relieved by the possible existence of a heavy lepton with a threshold not far from that of charm. The multiplicity crunch, as well as other crises for charm production, were diluted by the possible presence of another new particle which could decay into low multiplicity channels, allowing the charmed particles to balance this with high multiplicities. Furthermore, only $\frac{1}{4}$ of the cross section at 4.8 GeV need then be due to charm, rather than the ~$\frac{1}{2}$ assumed before: the limits on branching ratios into specific channels rise accordingly.

But now it is possible to look at the charged multiplicity on and off the bumps in the 4 GeV region. The data[34] show little or no change on passing through these bumps! The crunch would seem to be back - a reasonable estimate[35] of the number of charged particles per charmed particle decay is ≤ 2.3.

As can also be seen from previous data, the average momentum per charged particle must be less for final state hadrons resulting from the charmed quark contribution to R than from that from ordinary quarks. This follows already from the observation that the inclusive single particle distribution only changes for $x = 2p/E_{cm} \lesssim 0.5$ on crossing 4 GeV. Thus $<x>$ is less for any of the "new" physics than for the "old" physics in that region. If we assume each particle, charged or neutral, has the same average momentum, then a drop in the mean charged particle momentum means a jump of the total multiplicity. Thus it may be that charm is characterized by a greater total multiplicity than ordinary physics at the same $e^+ e^-$

energy, and if the charged multiplicity shows no increased multiplicity shows up in the neutrals. Whether
this possible change is because of neutrinos, or from
π°'s and/or γ's from $D^*{\to}D$ transitions[37] we do not know.
But at the rate experimental progress is being made, I
do not think this problem will survive until yet another
conference.   There is hope for proper "psychotherapy"
by then.

REFERENCES

1. J. E. Augustin et al., Phys. Rev. Letters 34, 764
    (1975).

2. W. Tanenbaum, invited talk at this conference; see
    also J. Siegrist et al., SLAC-PUB-1717, 1976 (un-
    published).

3. M. L. Perl et al., Phys. Rev. Letters 35, 1489 (1975).

4. F. J. Gilman, in Theories and Experiments in High
    Energy Physics, A. Perlmutter and S. Widmayer, eds.
    (Plenum Press, New York, 1975), p. 29.

5. B. J. Bjorken and S. L. Glashow, Phys. Letters 11,
    255 (1964).

6. S. L. Glashow, J. Illiopoulos, and L. Maiani, Phys.
    Rev. D2, 1285 (1970).

7. G. Zweig, CERN preprints TH. 401 and TH. 412, 1964
    (unpublished); also J. Iizuka, Suppi, Prog. Theor.
    Phys. 37-38, 21 (1966).

8. H. Oberlack, invited talk at this conference.

9. For a summary and references on the $C=+$ states found
    at DESY and SLAC, see the talks of B. H. Wiik and G.
    J. Feldman, respectively, in Proceedings of the 1975
    International Symposium on Lepton and Photon Inter-
    actions at High Energies, W. T. Kirk, editor (Stanford
    Linear Accelerator Center, Stanford, 1976), pps. 69
    and 39.

10. M. Bander et al., UC-Irvine preprint No. 75-54, 1975
    (unpublished). C. Rosenzwieg, University of Pitts-
    burgh preprint PITT-158, 1975 (unpublished).

11. Recall that the SPEAR scan for narrow resonances
    possesses a sensitivity which is expressed in terms
    of the area (in say nb-MeV) under the possible re-
    sonance peak which would lead to a detectable state.
    This area is proportional to the e^+e^- decay width

divided by M_R^2.

12. The full scan up to ~ 7.6 GeV and references are
     found in R. F. Schwitters, Proceedings of the 1975
     International Symposium on Lepton and Photon Inter-
     actions at High Energy, W. T. Kirk, editor (Stanford
     Linear Accelerator Center, Stanford, 1976), p. 5.

13. A. M. Boyarski et al., Phys. Rev. Letters 34, 1357
     (1975).

14. B. Jean-Marie et al., Phys. Rev. Letters 36, 291
     (1976).

15. F. J. Gilman, invited talk in High Energy Physics and
     Nuclear Structure-1975, D. E. Nagle, R. L. Burman,
     B. G. Storms, A. S. Goldhaber, and C. K. Hargrave,
     eds. (American Institute of Physics, New York, 1975),
     AIP Conference Proceedings No. 26, p. 331.

16. See the photon spectra in A. D. Liberman, Proceedings
     of the 1975 International Symposium on Lepton and
     Photon Interactions at High Energy, W. T. Kirk,
     editor (Stanford Linear Accelerator Center, Stanford,
     1976), p. 55.

17. Here K*≡K*(890), a member of the nonet of vector
     mesons, while K**≡K*(1420), a member of the nonet of
     tensor mesons.

18. The experimental situation is reviewed in G. S. Abrams
     Proceedings of the 1975 International Symposium on
     Lepton and Photon Interactions at High Energy, W. T.
     Kirk, editor (Stanford Linear Accelerator Center,
     Stanford, 1976), p. 25.

19. S. Rudaz, Cornell preprint CLNS-324, 1975 (unpublish-
     ed). Recall also that $\psi \to \pi\rho$ is known (REf. 14) to be
     overwhelmingly "direct" rather than electromagnetic
     in origin.

20. W. Tanenbaum et al., SLAC-PUB-1969, 1975 (unpublished)

21. In the usual spirit of vector dominance for a process on the photon-mass-shell, one uses the γ-ψ coupling as measured on the ψ-mass-shell in $\psi \to e^+ e^-$. Similarly, for ψ' vector dominance, one employs $\psi' \to e^+ e^-$.

22. The contributions from the "light" ρ, ω, and ϕ are negligible. Note that the additional heavy vector meson contributions which were omitted would have to cancel the calculated contributions from the ψ or ψ' almost completely to reproduce the experimental data.

23. SLAC-Wisconsin collaboration (private communication).

24. Technically, in the charmonium picture there is an additional dependence on the strong interaction (gluon-quark) coupling constant for the "direct" hadronic decays. However, as this change is only logarithmic, the coupling only changes slightly between 3.1 and 3.7 GeV.

25. See A. Liberman, Ref. 16, and G. J. Feldman, invited talk at the Palermo Conference, June 23-28, 1975 and SLAC-PUB-1624, 1975 (unpublished).

26. H. Harari, Weizmann Institute preprint WIS-75/39, 1975 (unpublished). See also in this connection C. Rosenweig, University of Pittsburgh preprint PITT-156, 1975 (unpublished) and Chan Hong-Mo et al., Rutherford Laboratory preprints RL-75-177 and RL-75-192, 1975 (unpublished).

27. See, for example, T. Appelquist and H. D. Politzer, Phys. Rev. Letters 34, 43 (1975).

28. G. J. Feldman, invited talk at the Irvine Conference, December, 1975 (unpublished).

29. A. Benvenuti et al., Phys. Rev. Letters 34, 419 (1975).

30. B. Barish in Proceedings of La Physique du Neutrino a Haute Energie (Ecole Polytechnique, Paris, 1975), p. 131.

31. Gargamelle collaboration, CERN preprint, 1975 (un-
    published), and H. Deden et al., Phys. Letters 58B,
    361 (1975).

32. J. von Krogh et al., University of Wisconsin preprint
    1975 (unpublished).

33. A. M. Boyarski et al., Phys. Rev. Letters 35, 196
    (1975).

34. See R. F. Schwitters, Ref. 12.

35. This comes from calculating the largest possible
    multiplicity due to a "bump" and assuming it all aris
    from the weak decays of two charmed particles. Since
    "ordinary" pions, etas, etc. will generally be pro-
    duced in the same event as charmed particles, this
    number is certainly an upper limit, given our assump-
    tions.

36. Inasmuch as any change in the total multiplicity is
    small, and a small change in the charged multiplicity
    on a bump in R cannot be ruled out yet, the increase
    in neutrals must be considered as very tentative.

37. See, for example, S. Nussinov, Institute for Advanced
    Study preprint COO2220-54, 1975 (unpublished).

QUARK-TASTING WITH NEUTRINOS*

A. De Rújula

The Physics Laboratories

Harvard University, Cambridge, Mass.   02138

ABSTRACT

I review the current situation in neutrino physics
in the context of unified gauge theories, with room for
heavy quarks that carry new flavor quantum numbers. The
y-anomaly is strong evidence for the existence of right-
handed currents and perhaps beauty. Dileptons are evidence for charm. The case for truth is less compelling.
I study equal sign dileptons and their implications on
a promising field of multilepton physics. Neutral
currents, the hard-fact model and madness (nondiagonal
neutral currents) are also among the topics discussed.

I.   INTRODUCTION

My task is to review the recent progress in understanding high energy neutrino phenomena. The task is
somewhat simplified by the fact that many of the striking new facts had been anticipated by theory (opposite
sign prompt dileptons,[1] violations of scaling and

*Work supported in part by the National Science Foundation
  under Grant No. MPS75-20427.

charge symmetry,[2,3] y-anomalies,[4] high hadron mass
anomalies,[2] deviations from the Adler and Llewellyn-
Smith sum rules[5,6]...). The task is made difficult by
the inherent limited statistics of neutrino experiments.
This limitation often induces theorists to use the data
the way drunkards use lampposts (for support rather
than illumination). This problem I can only partially
overcome.

I discuss the data in the context of the quark-
parton model.[7] The model offers a simple and consistent
description of deep inelastic inclusive charged lepton
scattering and, in its naive version where the nucleon
consists mainly of three "valence" quarks, the model
adequately predicts the inclusive differential cross
section for $\nu(\bar{\nu})$ scattering at intermediate energies[7,4]
(2 GeV $\lesssim E_\nu \lesssim$ 30 GeV). The coupling of intermediate
vector bosons to quarks is treated as pointlike, neutrino data are presumably not yet accurate enough to
detect the small logarithmic deviations from exact
scaling predicted by asymptotically free field theories[8]
and perhaps already observed in charged lepton
scattering.[9] If new hadronic degrees of freedom associated with heavy quarks exist, neutrinos may singly
excite them via large calculable pointlike couplings.
The large masses of the new particles play a nontrivial
role in the theoretical description of the "rescaling"
transition from the intermediate energy scaling region
to a higher energy scaling domain, where the effects of
the heavy masses also become negligible.[10] As a matter
of fact, the effects of heavy quarks in the choice of
the "right" scaling variable[11] turn out to be essential
to an overall understanding of inclusive high energy
neutrino scattering.[12]

To display the notation and illustrate specific points, I refer to two types of unified models of the weak and electromagnetic interactions based on the gauge group SU(2) ⊠ U(1):[13] the "standard" model, and the "strictly vectorlike" model. Throughout, the strong interactions are conventional quantum chromodynamics (QCD) based on color SU(3). In the standard model the weak current is the purely left-handed GIM current.[14] There are four quark flavors (p, n, λ, c) sitting in two weak SU(2) doublets:

$$\begin{pmatrix} p \\ n_\theta \end{pmatrix}_L ; \begin{pmatrix} c \\ \lambda_\theta \end{pmatrix}_L , \begin{cases} n_\theta = n \cos\theta + \lambda \sin\theta \\ \lambda_\theta = -n \sin\theta + \lambda \cos\theta \end{cases} \qquad (1)$$

In the strictly vector models every quark sits in a right- and a left-handed doublet. A minimum of six quarks is necessary and the doublets are:

$$\begin{pmatrix} p \\ n_\theta \end{pmatrix}_L ; \begin{pmatrix} c \\ \lambda_{\theta,\phi} \end{pmatrix}_L ; \begin{pmatrix} t \\ b_{\theta,\phi} \end{pmatrix}_L ; \begin{cases} \lambda_{\theta,\phi} = \lambda_\theta \cos\phi + b \sin\phi \\ b_{\theta,\phi} \perp n_\theta, \lambda_{\theta,\phi} \end{cases} ; \qquad (2a)$$

$$\begin{pmatrix} p \\ b \end{pmatrix}_R ; \begin{pmatrix} c \\ n_\psi \end{pmatrix}_R ; \begin{pmatrix} t \\ \lambda_\psi \end{pmatrix}_R ; \begin{cases} n_\psi = n \cos\psi + \lambda \sin\psi \\ \lambda_\psi = -n \sin\psi + \lambda \cos\psi \end{cases} \qquad (2b)$$

where t and b stand for truth and beauty. What the value of the angle ψ is reflects two orthogonal points of view. The CUNY-Harvard[15] view is $\cos\psi \sim 1$, while Caltech-Princeton[16] advertise $\cos\psi \sim 0$. In strictly vectorlike theories the neutral currents are purely vector. Pre-

liminary experimental evidence against this possibility
may exist.[17] Were this evidence to survive further
analysis, one may retreat into partially vector theories[1?]
where some of the quarks remain in left- or right-handed
singlets.

The distortions of scaling due to heavy thresholds
are discussed in Section II. In Section III, I analyze
the inclusive single muon data. The antineutrino
y-anomaly[19] will inescapably force us to take the existence of right-handed currents and perhaps beauty very
seriously indeed. Opposite charge dileptons are
discussed in Section IV. They are evidence for a new
quantum number, perhaps charm and/or truth. In Section
V, I discuss equal sign dileptons, whose existence har-
bingers a fascinating field of multi-prompt-lepton
physics. In Section VI I briefly review the neutral
current situation.

II.   SCALING AND RESCALING IN NEUTRINO SCATTERING

In Fig. 1a, I have plotted the usual (Q^2, ν) plane
of the variables that describe inclusive neutrino
scattering. The customary definitions of variables can
be found in Fig. 1b. In theories with heavy quarks there
are fixed W-thresholds where the real particles with the
corresponding quantum numbers start to be produced. The
thresholds of six quark theories appear in the figure
in an order dictated by my unjustified prejudice (to be
given some support in the subsequent analysis of data).
The effect of these thresholds on inclusive stattering
will be governed by one or more mass scales, referred
to in the following as "heavy quark masses, m_q." It is
possible to justify in QCD that the trade of heavy real

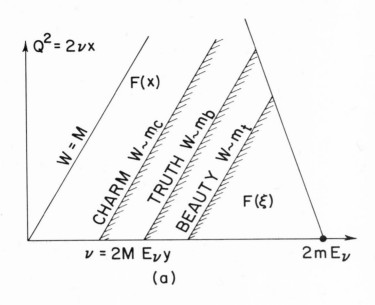

(a)

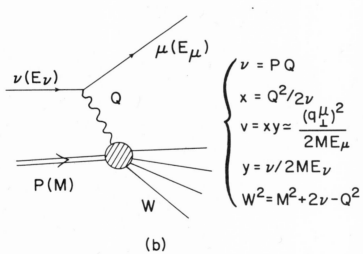

(b)

Figure 1   a)   The Kinematical (Q^2-ν) plot with fixed W
               thresholds.
          b)   The conventional kinematical definitions
               in inclusive lepton scattering.

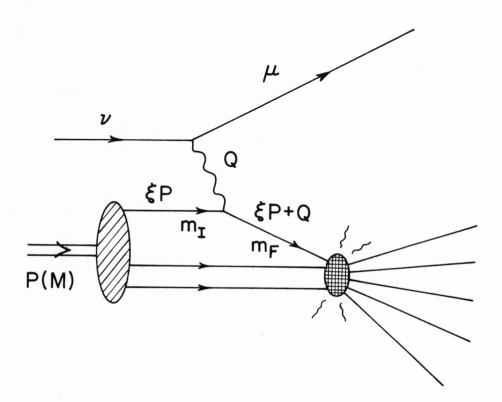

Figure 2   Definition of the Georgi-Politzer juvenile
scaling variable ξ.

masses for quark masses makes sense in this
context.[10,11,20,21] For $W \gtrsim 2$ GeV, $Q^2 \gtrsim 1$ GeV and below
eventual thresholds, electron and neutrino scattering
are known to satisfy approximate Bjorken scaling or
"improved" versions of it.[22] From dimensional analysis
we expect scaling to be recovered well above threshold.
The way the transition takes place is crucial to the
analysis and understanding of the neutrino data.[11,12]

To illustrate the maximum threshold effects that
one may conceivably expect, the authors of Reference 4
considered "fast" rescaling, in which all quark masses
(including the heavy ones) are neglected and the
structure functions get an additional contribution
$F(x)\theta(W-W_{th})$ above threshold. In this approximation,
as in the orginal parton model, x is both the fraction
of longitudinal momentum of the struck quark (in the
∞-momentum frame) and the experimentally measured
quantity $x = Q^2/2\nu$. To emphasize that this could only
be an overestimate, the same authors studied a version
of "slow" rescaling, where the x variable is shifted
to the new threshold $x' = Q^2/(\nu-\nu_{th})$ and the addition
is $F(x')\theta(W-W_{th})$.
More recently Georgi and Politzer[11,20] have derived
the form of scaling and rescaling to be expected in the
framework of QCD. Let the initial (final) quark have
mass $m_I(m_F)$ and let ξ be the fraction of four-momentum
carried by the struck quark, as in Fig. 2. The meaning
of x is unchanged, x is the experimentally determined
ratio $Q^2/2\nu$. Following Georgi and Politzer, define

$$\xi \equiv \frac{Q'^2}{2M\nu} \frac{2}{(1 + (1+Q'^2/\nu^2))^{\frac{1}{2}}} \qquad , \qquad (3a)$$

$$2Q'^2 \equiv Q^2 + m_F^2 - m_I^2 + \sqrt{Q^4 + 2Q^2(m_F^2 + m_I^2) + (m_F^2 - m_I^2)^2} \quad ,$$

$$(3b)$$

where M is the proton mass. The claim is that one can derive ξ-scaling laws from a thorough analysis of the short-distance operator product expansion. An example is[11]

$$\frac{\sqrt{1 + 4x^2 \frac{M^2}{Q^2}}}{2x^2} \left\{ 6xW_1 - \left(1 + \frac{4x^2 M^2}{Q^2}\right) \frac{\nu W_2}{M^2}\right\} = F(\xi) \qquad (4)$$

and similar but more complicated laws for W_3 and the "longitudinal" combination of W_1 and W_2. Here only the effects of masses are being considered, the "usual" logarithmic deviations from scaling are neglected. The known x-scaling laws are recovered as $Q^2 \gg m_I^2$, m_F^2, M^2. For m_I^2, m_F^2 negligible relative to Q^2, the real situation in electron or muon scattering off a target (protons or neutrons) that allegedly contains only a small contamination of heavy quarks, the "right" GP-scaling variable is, to leading order in M^2/Q^2

$$\xi \simeq x/(1 + x^2 \frac{M^2}{Q^2}); \quad (Q^2 \geq 0) \quad . \qquad (5)$$

This is quite close to the successful phenomenological variable of Bloom and Gilman,[22]

$$\tilde{x} = x/(1 + x \frac{M^2}{Q^2}), \qquad (6)$$

particularly at large x, where $\tilde{x}$ does most of its job at describing the moderately low Q^2 and W^2 data in terms of only one variable. Georgi and Politzer have thus gone one step ahead towards a field-theoretic understanding of juvenile scaling.

Back to neutrino scattering and contrary to the
electroproduction case, heavy quarks can be made, and
are important. On a nuclear target a light quark
(p, n, λ; $m_I^2/Q^2 \sim 0$) may be turned into a heavy quark
for which m_F^2/Q^2 is not negligible until well above
threshold. To leading order in M^2/Q^2 and m_F^2/Q^2 the
"right" scaling variable is:

$$\xi \simeq \left(x + \frac{m_F^2 - m_I^2 x^2}{2MEy} \right) \left(1 - \frac{x^2 M^2}{Q^2} \right) \qquad (7a)$$

$$\simeq x + \frac{m_F^2}{2MEy} \equiv x + \frac{y_{th}}{y} \qquad . \qquad (7b)$$

The second approximation above, Eq. (7b), is justified by the smallness of (x^2) (in electron scattering
off deuterium $(x^2) \sim 0.085$). Equation (7b) has been
guessed by many[23] in the context of the naive parton
model. It is the result of setting all masses except
m_F to zero in Fig. 2, and imposing $(\xi P+Q)^2 = m_F^2$, i.e.
the "mass shell condition for the final heavy quark."
It is satisfactory that the parton model result can
be derived in a context where a heavy quark mass parameter allegedly makes sense.[21]

Equation (7b) tells us that, if x is the measured
quantity in a neutrino event that produced a heavy
quark, ξ, the fractional four-momentum of the parent
quark was bigger. Since the distribution $F(\xi)$ vanishes
fast as $\xi \to 1$, the effect is to make cross sections at
finite energies much smaller than estimated with fast
rescaling: GP rescaling is very slow.

Above a threshold associated with quark mass m_F, the

inclusive $\nu(\bar{\nu})$ cross sections get additional contributions:

$$\frac{\pi \Delta d\sigma}{G^2 MEdxdy} = \left\{ 1 - y + \frac{Mxy}{2E} + \frac{x}{\xi}\left(\frac{y^2}{2} \pm \left[y - \frac{y^2}{2}\right]\right)\right\} F_2(\xi)\theta(1-\xi),$$

$$(8)$$

where $F_2(\xi)$ is the customary scaling function for the struck quark. The x/ξ factor appears because the GP-scaling version of the Callan-Gross relation reads $F_2 = -\xi F_3$. The upper sign in Eq. (8) corresponds to left-handed currents, neutrino beam, quark target. Each trade of left for right, ν for $\bar{\nu}$ or quark for antiquark flips the sign. In Eq. (8) the "longitudinal" structure function has been neglected, since it is expected to vanish as a small multiple of M^2/Q^2.[11]

To develop a feeling for Eq. (8), I have plotted in Fig. 3[24] the differential cross sections $d\sigma/dy$ and $d\sigma/dx$ for antineutrino induced events on an approximately isoscalar target. I have added to the naive valence quark model the effects of a righthanded current $(p,b)_R$. This is done for different energies above the threshold energy $E_{th} \sim m_b^2/2M$. The effects on the y distribution are remarkable. Not so the effects on the x distribution, (x) goes down and then up with energy, but only a few percent. In Fig. 4,[24] I have done the same exercise for neutrinos, with an extra right-handed current $(t,n)_R$. The effects are much less dramatic than for antineutrinos. In Fig. 5,[24] I show the effects of GP-rescaling on the distribution of invariant mass of the extra new hadrons, for the antineutrino case. Relative to fast rescaling estimates, GP-rescaling somewhat shifts the $d\sigma/dW$ distribution towards higher W. I have deliberately avoided

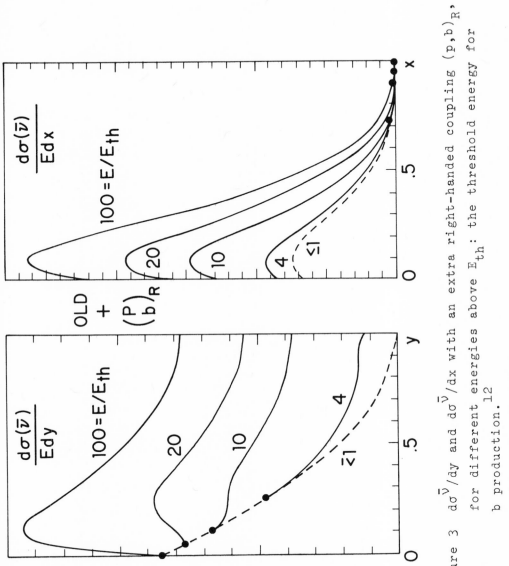

Figure 3 $d\sigma^{\bar{\nu}}/dy$ and $d\sigma^{\bar{\nu}}/dx$ with an extra right-handed coupling $(p,b)_R$, for different energies above E_{th}: the threshold energy for b production.[12]

showing distributions based on struck sea quarks. We
know little about their ξ-dependence, and often the
predictions based on the sea involve as much weak
interaction theory as strong interaction guess work.

III.   INCLUSIVE SINGLE MUON DATA AND THE y-ANOMALY

The recent history of high energy neutrino physics
is a rosary of expected surprises. The total cross
sections were found to rise linearly with energy, a
scaling prediction. The ratio R of charged current
cross sections

$$R = \frac{\sigma(\bar{\nu}N \rightarrow \mu^+ + \ldots)}{\sigma(\nu N \rightarrow \mu^- + \ldots)} \qquad (9)$$

was found to be almost as small as 1/3, the naive valence
quark model value. In a scaling and charge symmetric
theory $R \simeq 1/3$ implies a $d\sigma/dy$ distribution flat for
neutrinos and $\sim(1-y)^2$ for antineutrinos,[2,3] the observed
shape at intermediate energies. The magnitude of the
neutrino cross section agrees with the prediction based
on electron scattering as obtained, say, in a model with
fractionally charged quarks. Neutral currents were
found, at the level predicted by the Weinberg-Salam
model with the simplest mass-generation mechanism for
the intermediate vector boson masses.

As if the above was not enough, fancier effects were
foreseen, due to the anticipated production at high
energies of new forms of hadronic matter. The $d\sigma/dy$
distribution for antineutrinos in particular, was recognized as being most sensitive to new channels becoming kinematically accessible.[4] The effect should
be an energy dependent high y anomaly, should make $(y)_{\bar{\nu}}$

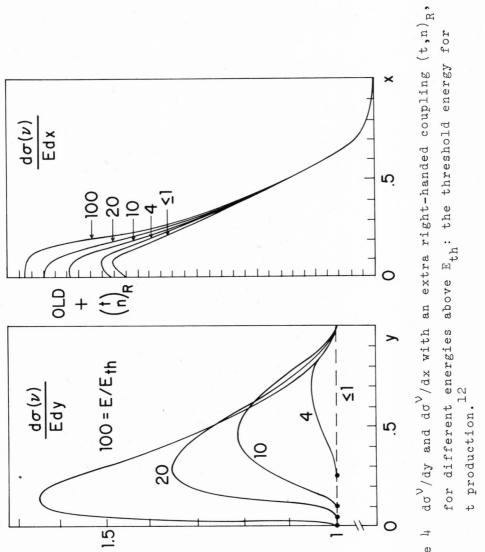

Figure 4 $d\sigma^\nu/dy$ and $d\sigma^\nu/dx$ with an extra right-handed coupling $(t,n)_R$, for different energies above E_{th}: the threshold energy for t production.[12]

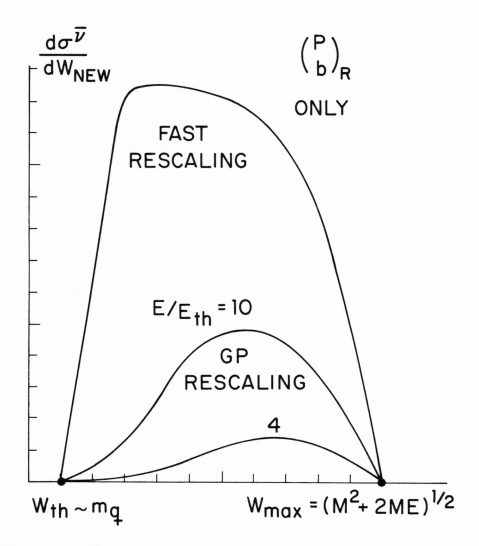

Figure 5 Shape of the hadron mass distribution of the extra events produced by monochromatic anti-neutrinos if a coupling of the form $(p,b)_R$ exists.

increase as a function of energy and be dramatic only
if the p and/or n quarks had right-handed couplings to
"fancy" heavy quarks.[4] This y anomaly has been a fact
for some time[25] and has further improved its cred-
ibility with more recent analysis.[19]

The data[25] on $(y)_{\bar{\nu}}$ is shown in Fig. 6. If scaling
was a correct assumption there would be no energy dependence of the scaling variable (y). Figure 6 is
the strongest evidence we have for a breakdown of
scaling in inclusive neutrino scattering.

Let me digress to prove that the standard four-quark
model cannot possibly describe the observed energy de-
pendence of $(y)_{\bar{\nu}}$. Let Q, $\bar{Q}$, Λ, $\bar{\Lambda}$, etc. be the ξ-inte-
grated light quark distributions:

$$Q = \frac{P + N}{2} \quad , \qquad \bar{Q} = \frac{\bar{P} + \bar{N}}{2}, \quad \ldots \qquad (10a)$$

$$P = \int_0^1 2\xi p(\xi)d\xi\ldots \quad , \qquad (10b)$$

where the conventionally defined $p(\xi)$ is the probability
of finding a p-quark in the proton with fractional four-
momentum ξ in the infinite momentum frame, and analogously
for other quarks. Below heavy quark thresholds and on
approximately isoscalar targets:

$$\frac{\pi}{G^2ME} \frac{d\sigma^{\nu}}{dy} = Q + \bar{Q}(1-y)^2 \quad , \qquad (11a)$$

$$\frac{\pi}{G^2ME} \frac{d\sigma^{\bar{\nu}}}{dy} = Q(1-y)^2 + \bar{Q} \ . \qquad (11b)$$

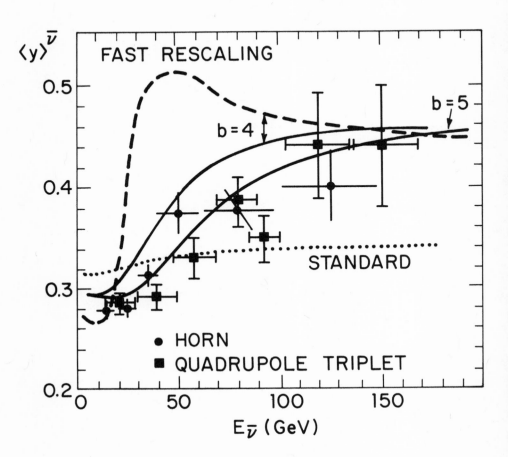

Figure 6   Experimental data for $(y)^{\bar{\nu}}$ as a function of
           energy. The dotted line is the prediction with
           fast rescaling in a model with the coupling
           $(p,b)_R$, for $m_b = 4$ GeV. Also shown (continuous
           lines) are GP-rescaling curves for $m_b = 4,5$ GeV
           in the same model. The dotted line is a best
           "fit" in the standard model.[12]

The fractional amount of antiquark "sea" is related to the total cross section ratio R

$$\alpha \equiv \frac{\bar{Q}}{Q} = \frac{3R - 1}{3 - R} \sim 0.05 \times (1 \pm 0.5) , \qquad (12)$$

where, for illustration purposes, I have used the low energy result R = 0.38 ± 0.02.[26] The small parameter α measures the deviations from absolute naiveté. The average $(y)_{\bar{\nu}}$ at low energies is predicted to be

$$(y)_{\bar{\nu}} = \frac{1}{4} \frac{1 + 6\alpha}{1 + 3\alpha} \sim 0.284 , \qquad (13)$$

not in disagreement with Fig. 6.[27] Well above charm threshold the standard model predicts an increase in cross sections which is sea-suppressed and/or Cabibbo-suppressed,

$$\frac{\pi}{G^2 ME} \frac{\Delta d\sigma^{\nu}}{dy} \rightarrow \sin^2 \theta_c Q + \Lambda , \qquad (14a)$$

$$\frac{\pi}{G^2 ME} \frac{\Delta d\sigma^{\bar{\nu}}}{dy} \rightarrow \bar{\Lambda} . \qquad (14b)$$

It seems reasonable to assume $\Lambda \approx \bar{\Lambda} \lesssim \bar{Q}$. The equality sign corresponds to an SU(3) symmetric sea. With this ansatz equations (11), (12) and (14) imply

$$\lim(y)_{\bar{\nu}} = \frac{1}{4} \frac{1 + 12\alpha}{1 + 6\alpha} \sim 0.31 , \qquad (15)$$

only $\sim$ 10% higher than the low energy result Eq. (13). Even for arbitrary α the maximum energy-dependent effect one can get is

$$\max \frac{(y)^{\bar{\nu}}, \text{ high E}}{(y)^{\bar{\nu}}, \text{ low E}} = 1.125 \quad (\alpha=1/6), \quad\quad (16)$$

much less than the data indicates.[28] The use of an
SU(4)-symmetric sea ($C=\bar{C}=\Lambda$) does not change the conclusion. Thus, the data in Fig. 6 is evidence for
something different or beyond the four quark model. The
figure shows Barnett's best "fit" for the standard model
($\alpha \sim 10\%$, $m_c \sim 1.5$ GeV) with the correct GP scaling
variable.

To understand $(y)_{\bar{\nu}}$ within the quark parton model it
is necessary to go as far as to assume that antineutrinos
couple to hadronic right-handed currents. Barnett has
calculated[12] the effects to be expected in models with
the right-handed coupling $(p,b)_R$ of Eq. (2b), with use
of the "right" GP-rescaling variable of Eq. (7b). Results for $m_b = 4,5$ GeV are shown in Fig. 6. Also
shown is an attempt to fit the data with a "fast" rescaling variable[29] ($\xi=x$). Fast scaling is seen to be
too abrupt to reproduce the apparent gentle behavior of
the data. An additional left-handed coupling $(p,l)_L$
would not sufficiently enhance high values of y to
accommodate the data. The conclusion that we are seeing
a right-handed current seems inescapable. Moreover, the
current must couple to the p and/ or n principal quarks.
That this current is $(p,b)_R$ as advertized by vector
theories, rather than $(n,g)_R$ or both cannot be decided
on the basis of the present experiment.[30]

Quite obviously a new quark implies a new res-
onance(s) with the quantum numbers of the photon. Presumably this resonance(s) is not $\psi(\psi')$, if these are

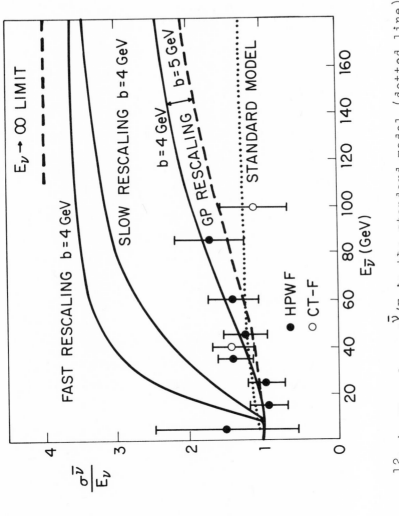

Figure 7^{12} a)   The slope $\sigma^{\bar{\nu}}/E$ in the standard model (dotted line) and in models with a coupling $(p,b)_R$. Fast, slow and GP-rescaling curves are shown.

associated with the $\sigma(e^+e^-)$ threshold at $\sim$ 4 GeV, in
which case the ψ constituent quark mass, in the same
sense used here, is $\sim$ 2 GeV.[10] Since m_b is $\sim$ 5 ± 1 GeV,
one naively expects a new narrow resonance or two at
9 ± 2 GeV, just right for the new e^+e^- machines to be
employable.[31] The very heavy mass of the new quark has
important consequences in antineutrino induced dimuons,
that I discuss in the next chapter.

     The large size of the y-anomaly forces us to ask
whether we should not have seen large effects somewhere
else in antineutrino data. The answer is negative for
the slope $\sigma^{\bar{\nu}}/E$ of the total cross section. This is
seen in Fig. 7a. For fast or slow rescaling (in the
sense of Ref. 4) there should be an observable effect,
but not (within errors) for the very slow GP-rescaling.[12]
Surprisingly enough, even at E $\sim$ 200 GeV and for
m_b $\sim$ 4 GeV the slope is only doubled (the high energy
limit is four times the low energy result). A large
y-anomaly entails an excess of events at large hadron
masses [$W=M^2 + 2MEy(1-x)$]. Such an excess above what
is expected for a $d\sigma/dy \sim (1-y)^2$ distribution has actually been seen (Fig. 8).[32] The high W threshold is
somewhere in the region W $\sim$ 4-5 GeV, in agreement with
the expectation from Barnett's fit to the y-anomaly.

     There is no hard evidence for the production of
new hadrons in the single muon neutrino induced events.
In the context of the standard or vector models we do
not expect very dramatic effects. Charm production is
sea- or Cabibbo-suppressed. Right-handed couplings add
$\sim (1-y)^2$ effects to $\sim$ 1 effects, the "signal/noise ratio"
is $\sim$ 1/9 as big as in $\bar{\nu}$ reactions, where the situation
is opposite.[4] The behavior of the slope σ^{ν}/E is
shown in Fig. 7b in the standard model (with $\sim$ 10% sea,

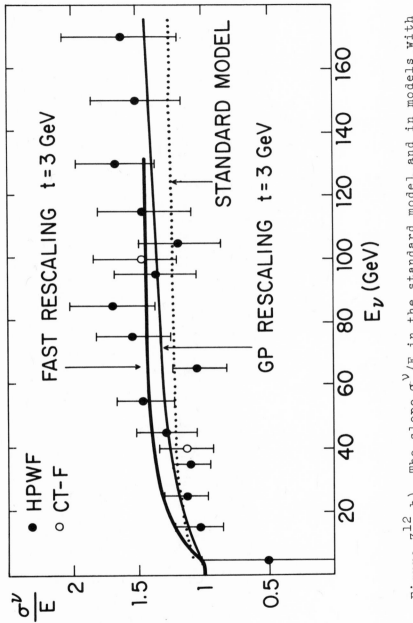

Figure 7[12 b)]   The slope σ^ν/E in the standard model and in models with an extra coupling $(t,n)_R$.

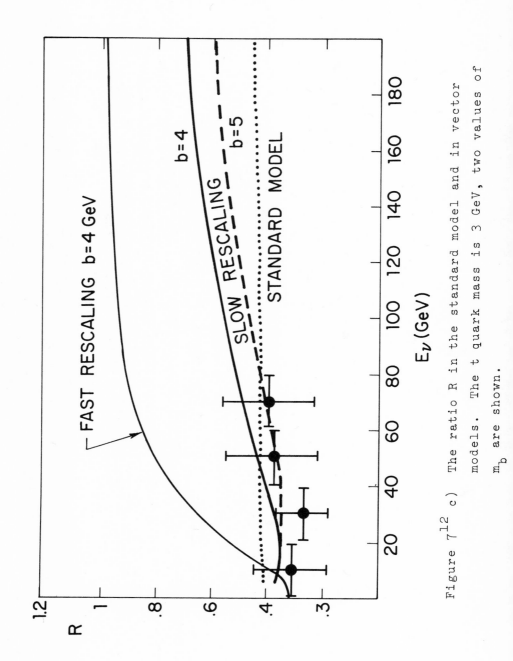

Figure 7[12] c) The ratio R in the standard model and in vector models. The t quark mass is 3 GeV, two values of m_b are shown.

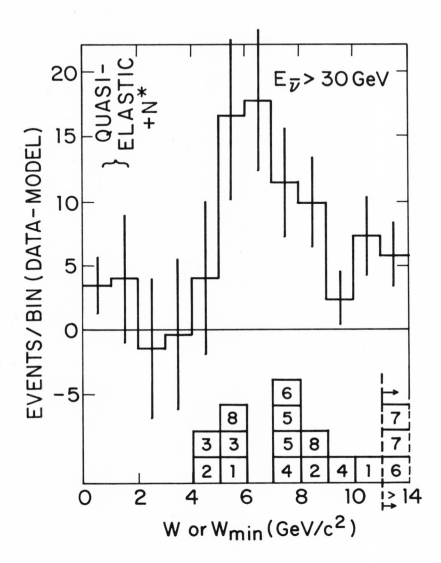

Figure 8   W   distribution of inclusive $\bar{\nu}$-induced events
with the conventional background subtracted
(top). W_{vis} distribution of ν-induced dimuon
events (bottom).   Data from the HPW-FNAL
collaboration.

$m_c \sim 1.5$ GeV) and in a vector model of the Caltech-Princeton variety (with $m_t \sim 3$ GeV), both with GP re-scaling.[12]

The behavior of $(y)_\nu$ as a function of energy is expected to be very smooth. Even with the extra coupling $(t,n)_R$, $(y)_\nu$ only changes from a low energy value $\sim 1/2$ to a high energy limit $\sim 7/16 \sim 0.44$. In inclusive ν scattering the most likely way new hadrons will show up is as structure in $d\sigma/dW$. Preliminary evidence for the expected effect is shown ·in Fig. 9.[33] There is also some evidence in $d\sigma^\nu/dx$ for the expected excess of low x events,[4] but it is not very compelling (Fig. 10).[32]

The ratio R of total cross sections is shown in Fig. 7c,[12] for the standard model ($m_c \sim 1.5$ GeV) and for a vectorlike model with $t \sim 3$ GeV, $b \sim 4,5$ GeV.

All of the above effects are expected to be accompanied by large deviations from charge symmetry,[2] an approximate property of the weak current only at low energies, where all terms but $(p,n_\theta)_L$ lie dormant. Scaling and charge symmetry imply, among other things, a definite shape of the ν minus $\bar{\nu}$ y distribution:

$$\frac{d\sigma^\nu}{dxdy} - \frac{d\sigma^{\bar{\nu}}}{dxdy} = \frac{G^2ME}{\pi} \, 2xF_3 y\left(1 - \frac{y}{2}\right) \, . \qquad (17)$$

The largest deviations from Eq. (17) should occur at low x. The HPW-FNAL data[26,32] for $x \leq 0.1$ are shown in Fig. 11.

A reanalysis of the y-anomaly has recently become available, with plots for y distributions at energies $E > 70$ GeV[19] (Fig. 12). From the published ν and $\bar{\nu}$ fluxes,[32] I estimate that the mean energy of events above 70 GeV is $(E) \sim 95$ GeV. This corresponds to

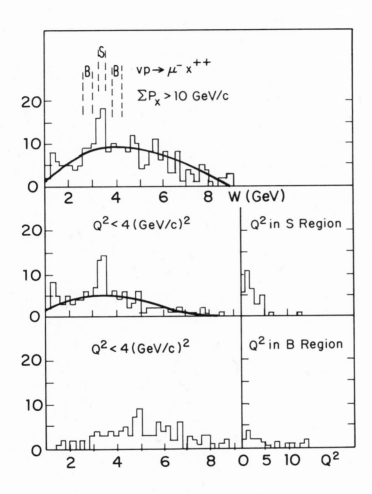

Figure 9   W distributions of ν-induced events.   FNAL, Michigan State Collaboration.[33]

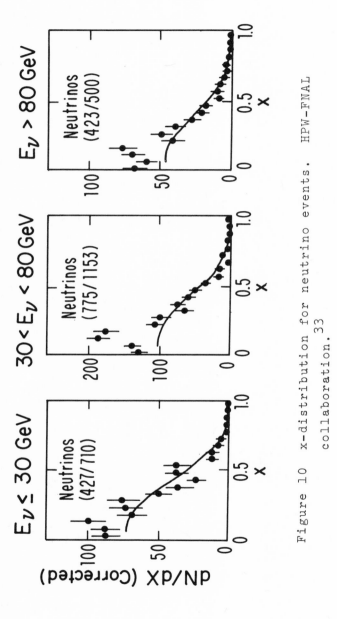

Figure 10   x-distribution for neutrino events.   HPW-FNAL
collaboration.[33]

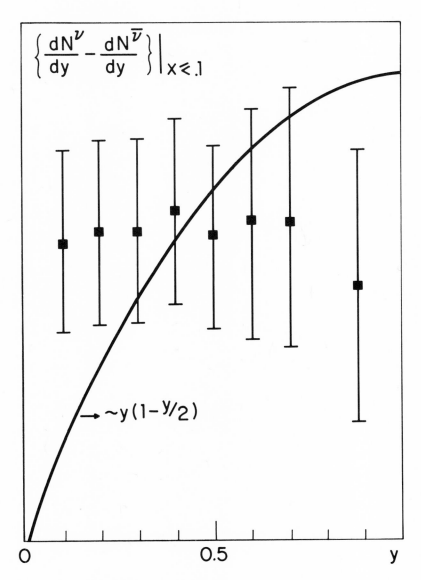

Figure 11   y-distribution of neutrinos minus antineutrinos
            for x≤ 0.1.   HPW-FNAL data.   The continuous
            curve is the prediction based on scaling and
            charge symmetry only.

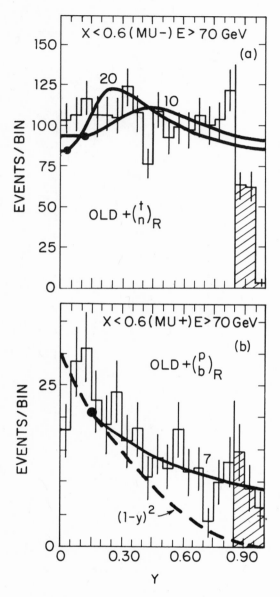

Figure 12 y distributions for ν and $\bar{\nu}$-induced events
with E > 70 GeV. Also shown are predictions
of models with right-handed currents. The
numerical labels stand for $\langle E \rangle / E_{th}$.

$(E)/E_{TH} \sim 7$ for $m_b \sim 5$ GeV (as estimated from Fig. 6). The predicted $d\sigma^{\bar{\nu}}/dy$ distribution, in a naive $+ (p,b)_R$ model with GP-rescaling, is plotted with the label "7" in Fig. 12b, and is consistent with the data. The $d\sigma^{\nu}/dy$ distribution, in a naive $+ (t,n)_R$ model is shown in Fig. 12a. The curve labelled "20" (for $(E)E/_{TH}=20$) corresponds to $m_t \sim 3$ GeV, the curve labelled "10" corresponds to $m_t \sim 4.2$ GeV. There is no evidence from this plot for or against a t quark of mass $m_t > 3$ GeV. A lighter t quark would give disagreement with the data.

IV. OPPOSITE SIGN DILEPTONS

The GIM current $(c,\lambda_\theta)_L$ obeys a $\Delta C = \Delta Q$ rule. Charmed particles will decay some of the time (semi-) leptonically and above charm threshold we expect the following reactions to take place:[1,2]

$$\nu_\mu N \rightarrow \mu^- + c + \dots$$

$$(\text{fast}) \hookrightarrow \begin{cases} \mu^+ \nu_\mu \\ e^+ \nu_e \end{cases} + \dots \text{ (mainly with S=-1)} \quad , \quad (18a)$$

$$\bar{\nu}_\mu N \rightarrow \mu^+ + \bar{c} + \dots$$

$$(\text{fast}) \hookrightarrow \begin{cases} \bar{\mu} \ \bar{\nu}_\mu \\ e^- \bar{\nu}_e \end{cases} + \dots \text{ (mainly with S=+1)} \quad , \quad (18b)$$

where "fast" stands for a lifetime of $10^{-12} - 10^{-14}$ secs. Opposite sign dileptons are "right" in the sense of the $\Delta C = \Delta Q$ rule. Dozens of opposite charge dimuons have been seen at the FNAL experiments.[34] Six events of the type $\nu N \rightarrow \mu^- e^+ K_s^0 + \dots$ were recently announced[35] along with three $\nu N \rightarrow \mu^- e^+ (\Lambda/K_s^0) + \dots$[36] and four $\nu N \rightarrow \mu^- e^+ + \dots$[36].

The dimuon events are evidence for the production
and subsequent weak decay of new particles. Thanks to
the well-known bounds on $(E_+)/(E_-)$ developed by Pais
and Treiman for lepton-mediated events,[37] it is established that neutral heavy leptons cannot be the only
source of dimuons. The $\mu^- e^+$ events look embarassingly
nice to charm believers, but for the miraculous add-
iction to K_s^0 of the Wisconsin et al.[35] data. Except
for this improbable manifestation of strangeness,
<u>neutrino-induced dileptons are compatible with charm
production</u>.

In the case of neutrino-induced dimuons,[34] the
data are abundant enough to allow for an estimate to be
made of the mass of the particle whose weak decay gives
rise to the μ^+.[38] The argument is based on the reasonable assumption that the new heavy hadron has a momentum pointing approximately in the direction of the
neutrino-muon momentum transfer, that lies in the plane
defined by the incident ν and outgoing μ^-. (See Fig.
13a). For any "old" particle and up to an exponentially
decreasing transverse momentum, this would be the observed behavior. The transverse momentum $k_\perp^+$ of the μ^+
and other decay products relative to the (ν,μ^-) plane is
unaffected by the motion of the parent particle, and can
be calculated in its rest frame. A description of the
observed[34] $dN/dk_\perp^+$, based on the elementary three-body
decay $c \rightarrow \lambda\mu\nu$ of Fig. 13b is shown in Fig. 13c.[38] The
k, cutoff approximately corresponds to half the new
particle's mass and the result ($M \sim 2\text{-}3$ GeV) is not unexpected for a charmed particle, or for a particle containing one quark of the variety of which J/ψ allegedly
contains two.

The statistics on antineutrino induced dimuons is

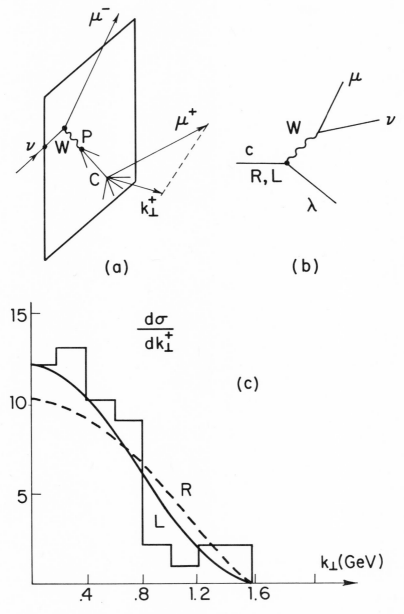

Figure 13 a) ν-production and leptonic decay of a charmed
          particle.
        b) "Elementary" charmed β-decay.
        c) Distribution in transverse momentum of the
           odd muon.[38]

not good enough for comparison of $d\sigma/dk_{\perp}^{-}$ with theoretical expectations. But, if the theoretical interpretation of the $(y)_{\bar{\nu}}$ anomaly via a $(p,b)_R$ current and the estimate $m_b \sim 4\text{-}5$ GeV are correct, the odd-muon transverse momentum distribution will not look the same for ν and $\bar{\nu}$ induced events. In Fig. 14 I have estimated $d\sigma(\bar{\nu} \rightarrow \mu^+\mu^-)/dk_{\perp}^{-}$ for different incident antineutrino energies, in a model with 5% $\bar{\lambda}$-quark sea, $m_c \sim 2$ GeV and $m_b \sim 4\text{-}5$ GeV. At low energies only charm is produced and the μ^- transverse momentum cuts at ~ 1.5 GeV. At higher energies beauty takes over and the distribution extends to ~ 2.5 GeV, a significantly higher cutoff than for ν-induced μ^+'s. We also expect the $d\sigma/dW_{vis}$ distribution of antineutrino-induced dimuons to be signnificantly shifted towards higher W_{vis}, relative to the neutrino-induced case ($W_{vis} \lesssim W$ in dimuon events, due to the missing outgoing meutrino).

The quoted[33] values for dimuon to single muon total cross sections are:

$$D_\nu \equiv \frac{\sigma(\nu \rightarrow \mu^-\mu^+)}{\sigma \nu \rightarrow \mu^-)} \sim (0.8 \pm 0.3)10^{-2}; \quad (E_{vis}) \sim 100 \text{ GeV}, \tag{19a}$$

$$D_{\bar{\nu}} \equiv \frac{\sigma(\nu \rightarrow \mu^+\mu^-)}{\sigma \nu \rightarrow \mu^+} \sim (2 \pm 1)10^{-2}; \quad (E_{vis}) \sim 90 \text{ GeV}. \tag{19b}$$

These numbers will turn out to be reasonable in the context of the standard model. But, since the standard model cannot accommodate the y-anomaly, I will also develop prejudices based on vector theories. In the standard model the cross sections rise to high energy limits:

$$\frac{\pi}{G^2 ME} \sigma(\nu \rightarrow \mu^-\mu^+) \rightarrow \{\sin^2\theta_c Q + \Lambda\}B \quad , \tag{20a}$$

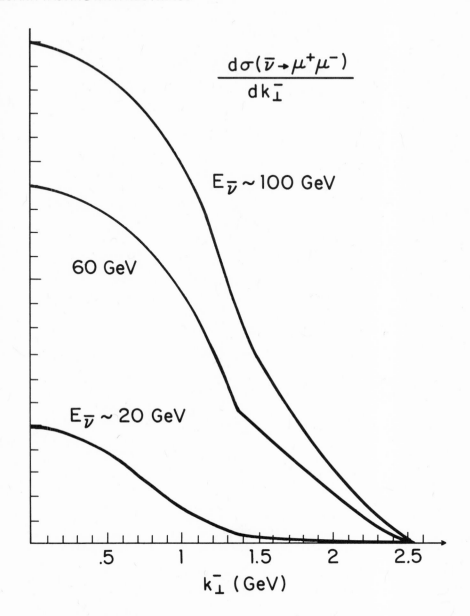

Figure 14   Anticipated transverse momentum distribution
            of the odd muon (μ^-) in antineutrino-induced
            dimuons. The model contains charm ($m_c \sim 2$
            GeV) and beauty ($m_b \sim 5$ GeV).

$$\frac{\pi}{G^2ME} \quad \sigma(\bar{\nu}\to\mu^+\mu^-) \to \{\bar{\Lambda}\}\bar{B} \ , \tag{20b}$$

where B and $\bar{B}$ are average (semi-) leptonic branching
ratios for the produced charmed and anticharmed states.
The rest of the notation is as in Eqs. (10), (11). The
ratios B and $\bar{B}$ may not be equal at low energies since
charmed baryons are only produced by neutrinos. At high
energies, however, a variety of charmed particles will
be ν- and $\bar{\nu}$-produced and one does not expect the aver-
ages B and $\bar{B}$ to be very different. In what follows
I take B = $\bar{B}$ for the sake of discussion.

The ratio of antineutrino to neutrino dimuon cross
sections is expected to rise faster to its asymptotic
value and contain fewer experimental and theoretical
uncertainties than the individual cross sections. In
the standard model

$$R_{\mu\bar{\mu}}\equiv\frac{\sigma(\bar{\nu}\to\mu^+\mu^-)}{\sigma(\nu\to\mu^-\mu^+)}=\begin{cases}\left(1 + \sin^2\theta_c\ \frac{3-R}{3R-1}\right)^{-1} & [SU(3)\ \text{symmetric sea}] \\ & \qquad\qquad\qquad (21a) \\ \left(1 + \frac{3}{4}\sin^2\theta_c\frac{3-R}{3R-1}\right) & [SU(4)\ \text{symmetric sea}] \\ & \qquad\qquad\qquad (21b) \\ 0 & [SU(2)\ \text{symmetric sea}], \\ & \qquad\qquad\qquad (21c)\end{cases}$$

where R is the single muon cross section ratio of Eq. (9).
Figure 15 is a plot of $R_{\mu\bar{\mu}}$ versus R. The prediction of
the standard model, which is not significantly different
for an SU(3) or SU(4) symmetric sea, is the dotted domain,
with R = 0.38 ± 0.02. In vector models both R and $R_{\mu\mu}$
are strongly energy dependent, but a relation of the
form $R_{\mu\mu}$ = f[R(E_ν)] still exists. Figure 15 shows the
prediction of a vector model and its high energy limit,
with the semileptonic branching ratios of charm, truth
and beauty set to be equal by fiat.

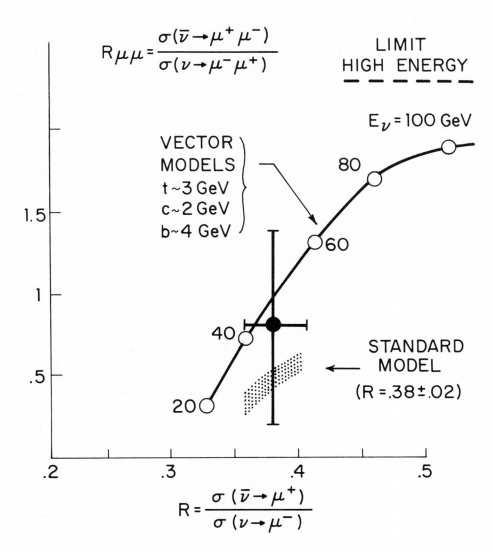

Figure 15   Dimuon versus single muon cross section ratios
             for ν̄ versus ν induced events. The standard
             model is the dotted domain. The prediction
             of a vector theory and the experimental point
             are also shown. The numbers in the vector
             prediction are neutrino energies in GeV.

In Fig. 16 I have plotted the ratios D_ν and $D_{\bar\nu}$ defined in Eq. (19) and the asymptotic predictions of the
standard model, with the ansatz of an SU(3) symmetric
sea. With a charmed quark mass $m_c \sim$ 1.5-2GeV and GP-
rescaling, the cross sections are within 10-20% of
their asymptotic value at E $\sim$ 40 GeV. It is seen from
the neutrino data in Fig. 15 that B $\sim$ 5-15%, a perfectly
reasonable semileptonic branching ratio for charmed
particles. This important result decreases only a
little if a right-handed coupling to a t quark exists,
provided $m_t \gtrsim$ 3 GeV.

We have found evidence for a right-handed coupling
of valence quarks to fancy (perhaps beautiful) quarks
of mass $\sim$ 4,5 GeV in the $(y)_{\bar\nu}$ anomaly, and evidence for
the production of new particles of mass 2-3 GeV, consistent with a coupling $(c,\lambda_\theta)_L$ to charmed quarks in the
neutrino induced dileptons. Is there evidence for anything else? The answer is yes, but very marginal.[12,29,34]
The indication comes from the y_{vis} distribution of the
μ^- in neutrino-induced dimuon events, shown in Fig. 17.
$(y_{vis} \lesssim y$ because of the missing neutrino). The dotted
"a" curve is the standard model, with only left-handed
couplings and GP-rescaling. The data seem to favor a
distribution which is closer to the $(1-y)^2$ shape expected asymptotically in a theory with right-handed
couplings. The curves "b" and "c" are obtained in a model
with an extra $(t,n)_R$ coupling, m_t = 3 GeV, and fast or
GP-rescaling, respectively.[12] There is inconclusive
evidence for truth in the data, with $m_t \sim$ 3 GeV. It
is amusing to notice that the existence of a 6-GeV resonance (truonium?) coupling to e^+e^- has been reported.[39]
The question arises of whether truth could be the only
new quantum number produced by neutrinos. The answer,

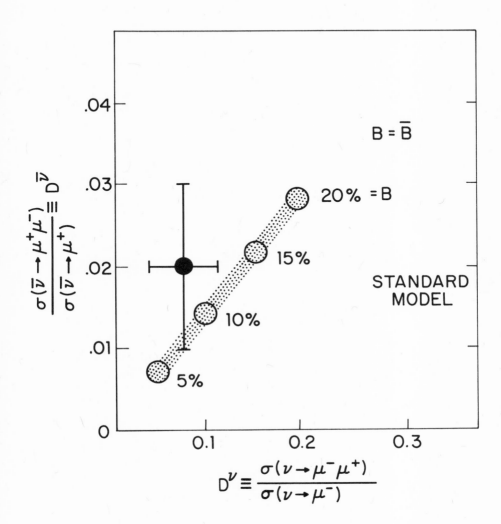

Figure 16 Antineutrino versus neutrino induced ratios of
           dimuon to single muon cross sections. The pre-
           diction of the standard model is shown, for
           different values of the branching ratio of
           charm to muons.

within the vector theories of Eq. (2), is negative.
This is because of the apparent overwhelming association
of the lower energy μ^-e^+ events with strangeness, a
signature for charm, not truth. The W_{vis} distribution
is not particularly sensitive to the handedness of
the coupling.[12] It is shown in Fig. 18,[12] where the
meaning of the labels is as before. The curve "c"
has fast rescaling, the W distribution gets shifted to
"b" by GP-rescaling.

In the standard model the x distribution of
dimuon events, in the high energy limit $\xi \to x$, is

$$\frac{d\sigma(\nu \to \mu^-\mu^+)}{dx[\mu^-]} \sim \sin^2\theta_c \frac{p(x) + n(x)}{2} + \lambda(x) , \quad (22a)$$

$$\frac{d\sigma(\nu \to \mu^+\mu^-)}{dx[\mu^+]} \sim \bar{\lambda}(x) . \quad (22b)$$

The conventional parton model wisdom is that $\lambda, \bar{\lambda}(x)$ will
be more peaked towards low x than $p, n(x)$. Thus the x
distribution for dimuons should be flatter in the
neutrino events than in the antineutrino events,[40] a
qualitative conclusion that should not be affected by
the difference between x and x_{vis} ($x_{vis} > x$ due to the
missing neutrino). This is the behavior of the data,
within pathetic statistics (Fig. 19). In a model with
a $(p,b)_R$ current, antineutrino-induced dimuons at high
energies have a valence quark-like x distribution. At
finite energies, however, the effect of slow rescaling
is to squeeze the distribution towards low x. In Fig.
19, I show Barnett's results[12] for the x distribution
in a variety of models. The dotted line (that in the
ν case coincides with the continuous line) is the
standard model, with $\lambda(x) \sim (1-x)^9$: a sea distribution

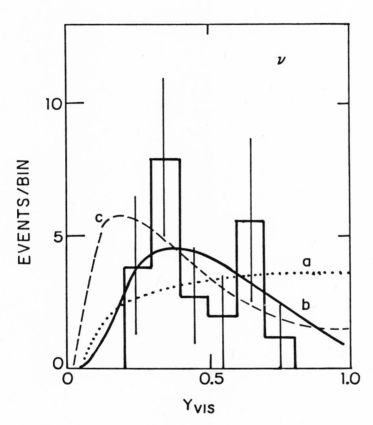

Figure 17 Y_{VIS} of the μ^- in neutrino induced dimuons.
See in the text the meaning of the curves.

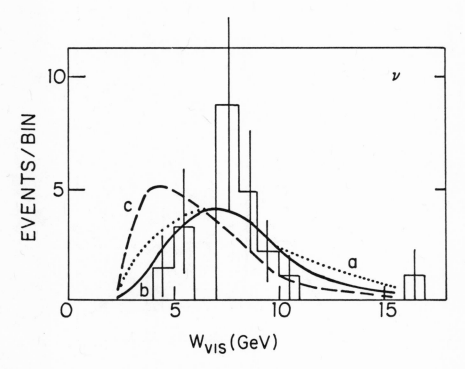

Figure 18 W_{VIS} distribution in neutrino induced dimuons.
See in text the meaning of the curves.

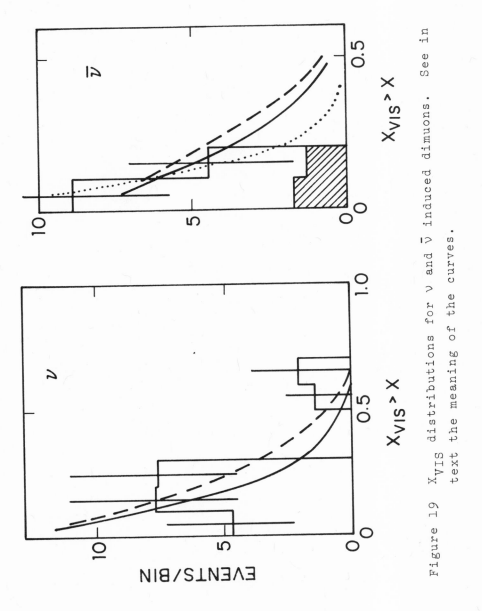

Figure 19 X_{VIS} distributions for ν and $\bar{\nu}$ induced dimuons. See in text the meaning of the curves.

very peaked towards low x. The other lines correspond to vector models with m_b = 4 GeV and m_t = 2,4 GeV (dashed and continuous curves, respectively). No strong conclusion can be drawn from the x distributions of the present data.

V.   EQUAL SIGN DILEPTONS AND MULTILEPTON PHYSICS

Events have been reported, apparently not mediated by conventional mechanisms, where neutrinos produced two equal sign muons.[34] While not entirely unexpected,[41] these events are one of the most interesting surprises of neutrino physics. The quoted cross sections are:[34]

$$\frac{\sigma(\nu \to \mu^- \mu^-)}{\sigma(\nu \to \mu^-)} \sim (0.8 \pm 0.5)10^{-3} \quad , \qquad (23a)$$

$$\frac{\sigma(\bar{\nu} \to \mu^+ \mu^+)}{\sigma(\bar{\nu} \to \mu^+)} \sim (0.8 \pm 0.7)10^{-2} \quad , \qquad (23b)$$

or, in terms of equal to opposite charge dimuon ratios:

$$\frac{\sigma(\nu \to \mu^- \mu^-)}{\sigma(\nu \to \mu^- \mu^+)} \sim 0.1 \pm 0.05 \qquad , \qquad (24a)$$

$$\frac{\sigma(\bar{\nu} \to \mu^+ \mu^+)}{\sigma(\bar{\nu} \to \mu^+ \mu^-)} \sim 0.4 \pm 0.3 \qquad . \qquad (24b)$$

Let me first discuss the events in the context of the standard model. Equal sign dimuons are "wrong" in the sense of the $\Delta C = \Delta Q$ rule and they can only be due to associated production of charm[41]

$$\nu N \to \mu^- c \bar{c} + \dots \qquad (25a)$$
$$ \hookrightarrow \mu^- \bar{\nu} + \dots \quad . \qquad (25b)$$

It was concluded in the previous section that neutrino-induced "right" dimuons yield an estimate of the branching ratio B = [C → μ +...]/[C → all] ∿ 5-15%. This, combined with (23a), implies that <u>associated charm production (25a), occurs in the standard model at the level of 1%</u>. Theorists would probably be equally divided on whether this is large of small, but it should be large eneough to make experimentalists ecstatic. The following are properties or consequences of the standard model interpretation of "wrong" dimuons:

i) The normalized cross sections should be similar for ν and ν̄:

$$\frac{\sigma(\nu \to \mu^- \mu^-)}{\sigma(\nu \to \mu^-)} \gtrsim \frac{\sigma(\bar{\nu} \to \mu^+ \mu^+)}{\sigma(\bar{\nu} \to \mu^+)} = B[\nu \to C\bar{C}\mu]B[C \to \mu] \quad . \quad (26)$$

The inequality is a possible consequence of the fact that neutrinos are better than antineutrinos at transmitting energy to the target.

ii) Both charmed particles can decay into leptons. Trimuons of specific signs are expected at the level of 10% of the "wrong" dimuons:

$$\frac{\sigma(\nu \to \mu^- \mu^- \mu^+)}{\sigma(\nu \to \mu^- \mu^-)} \sim \frac{\sigma(\bar{\nu} \to \mu^+ \mu^- \mu^+)}{\sigma(\bar{\nu} \to \mu^+ \mu^+)} = B[C \to \mu] \sim 5-15\% \quad . \quad (27)$$

The absence of trimuons at this level would kill the standard model interpretation.

iii) Trilepton and equal charge dilepton events should most of the time be accompanied by two particles of opposite strangeness. The following are interesting signatures for bubble chamber hunts:

$$\nu_\mu N \rightarrow \mu^- \ell^- S\bar{S} + \ldots \qquad (\sim 10^{-3} \sigma_{TOT}) \quad , \qquad (28a)$$

$$\nu_\mu N \rightarrow \mu^- \ell^- \ell^+ S\bar{S} + \ldots \qquad (\sim 10^{-4} \sigma_{TOT}) \quad , \qquad (28b)$$

where ℓ stands for e or μ, and S stands for a strange particle. Similar results hold for $\bar{\nu}_\mu$.

    iv) Neutral currents will also produce $C\bar{C}$ at the 1% level. This has amusing consequences, like

$$\nu_\mu N \rightarrow \nu_\mu C\bar{C} \rightarrow e^+ S\bar{S} + \ldots \qquad (10^{-3} - 10^{-4} \sigma_{TOT}) \quad ,$$

interesting only if the $\bar{\nu}_e$ contamination in the beam is small enough.

    v) Almost last but not least, the "observation" of $\nu N \rightarrow \mu^- C\bar{C}$ at the level of a percent sets a lower limit of the same order of magnitude for the electroproduction analog

$$\mu N \rightarrow \mu\ C\bar{C} + \ldots \qquad (\gtrsim 10^{-2} \sigma_{TOT}) \quad . \qquad (29)$$

This may be directly observed in emulsions in the form of short tracks. Trileptons and equal and opposite charge dileptons should also occur:

$$\mu^- N \rightarrow \mu^- \ell^+ S\bar{S} + \ldots \quad (10^{-3} \sigma_{TOT}) \qquad\qquad (30a)$$

$$\rightarrow \mu^- \ell^- S\bar{S} + \ldots \quad (10^{-3} \sigma_{TOT}) \qquad\qquad (30b)$$

$$\rightarrow \mu^- \ell^+ \ell^- S\bar{S} + \ldots \quad (10^{-4} \sigma_{TOT}) \quad . \qquad (30c)$$

Contrary to the neutrino case, equal and opposite charge dileptons must have similar cross sections.

    vi) No firm conclusion on the overall cross section

for the associated production of charm in hadron-hadron
collisions can be drawn. A branching ratio of 5-15% of
charm to leptons implies that 5-15% of hadron-induced
prompt leptons should be opposite sign dileptons.

Multilepton physics is particularly interesting
in vector models. Even if associated charm production
is negligible, wrong sign dileptons can occur via
D^0-$\bar{D}^0$ mixing[41,43] or cascades of quantum numbers. The
D^0-$\bar{D}^0$ system is the charmed pseudoscalar meson analog
of the K^0-$\bar{K}^0$ system. In the standard model D^0-$\bar{D}^0$ mixing
is Cabibbo suppressed, while D^0 decay is not (see Fig.
20a). Thus, the $\Delta C = 2$ mixing cannot be the source
of equal sign dileptons: the "wrong" decays $D^0 \to \mu^-$...
of a particle produced in a $\Delta C = +1$ process are suppressed
relative to the "right" decays $D^0 \to \mu^+$ ..., $D^0 \to$ hadrons
by a factor $\sim \tan^4 \theta_c \sim 10^{-3}$. In vector theories, however,
D^0-$\bar{D}^0$ mixing can be large. In models of the Caltech-
Princeton variety, for instance, the diagram of Fig. 20b
($\sim \sin^2 \phi m_b^2$) can produce strong mixing provided m_b is large
($\sin^2 \phi m_c^2$ is bounded by the K_L-K_S mass difference). The
following process is possible

$$\nu N \to \mu^- D^0 + ...$$
$$\qquad \rightarrow D_{S,L} \to \mu^- \bar{\nu} + \text{(mainly S = +1)} . \qquad (31)$$

A large but thinkable amount of mixing is necessary for
the "wrong" dileptons to originate via this process.
Whether this is the origin of equal sign dileptons can
be tested, since it implies very specific signatures
both in neutrino scattering and $e^+ e^-$ annihilation:

i) "Wrong" dileptons come with the "right"
strangeness (in the sense of the $\Delta S = \Delta Q$ rule), an
example of the reaction (31) is $\nu \to \mu^- e^- K^+$ ... Contrary-

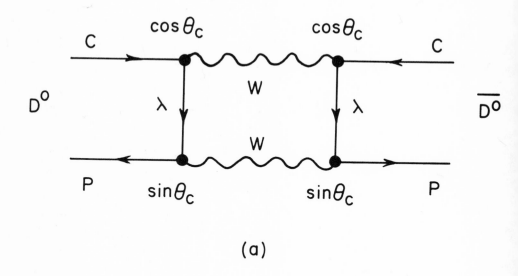

(a)

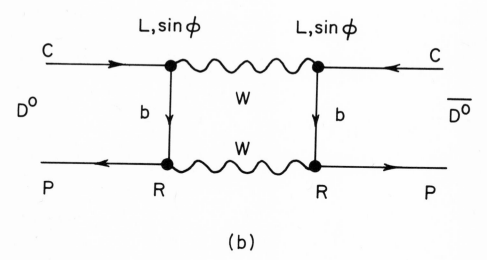

(b)

Figure 20   Examples of diagrams contributing to D^0-$\overline{D}^0$
             mixing.
     a)   In the standard model.
     b)   In vector models.

wise "right" dileptons have the "wrong" strangeness,
i.e.: $\nu \rightarrow \mu^- e^+ K^- \ldots$

    ii) In $e^+ e^-$ annihilation above charm threshold
$\Delta S = 2$ events will be found at the several percent
level:

$$e^+ e^- \rightarrow D^0 \; \bar{D}^0$$
$$\qquad\quad \rightarrow K^+ + \ldots$$
$$\qquad\quad \bar{D}^0 \rightarrow K^+ + \ldots \qquad\qquad (32)$$

At the perthousand level $\Delta L_c = \Delta S = 2$ events occur:

$$e^+ e^- \rightarrow D^0 \; \bar{D}^0$$
$$\qquad\quad \rightarrow K^+ \ell^- \bar{\nu} \ldots$$
$$\qquad\quad \bar{D}^0 \rightarrow K^+ \ell^- \bar{\nu} \ldots \qquad\qquad (33)$$

    In vector theories cascades of prompt decays can
also produce equal sign dileptons. An example is:

$$\bar{\nu} p \underset{R}{\rightarrow} \mu^+ b$$
$$\qquad \underset{L}{\rightarrow} t + \text{hadrons}$$
$$\qquad\quad \underset{R}{\rightarrow} n \, \mu^+ \nu \qquad\qquad (34)$$

With $m_b > m_t > m_c$ this process occurs on valence quarks
and with a cross section comparable to the observed one
only in $\bar{\nu}$-induced reactions. If one allows for cascades
or associated production of the quantum numbers of vector
theories, all sorts of multileptons are possible. An
example, illustrated at the top of Fig. 21, is "wrong
sign trileptons", $\nu_\mu \rightarrow \mu^- \ell^+ \ell^+$ (compare with $\nu_\mu \rightarrow \mu^- \ell^- \ell^+$,
Eq. (27), the only possible trilepton charge assignment
predicted by the standard model). More complicated
cascades can lead anywhere up to hexalepton events

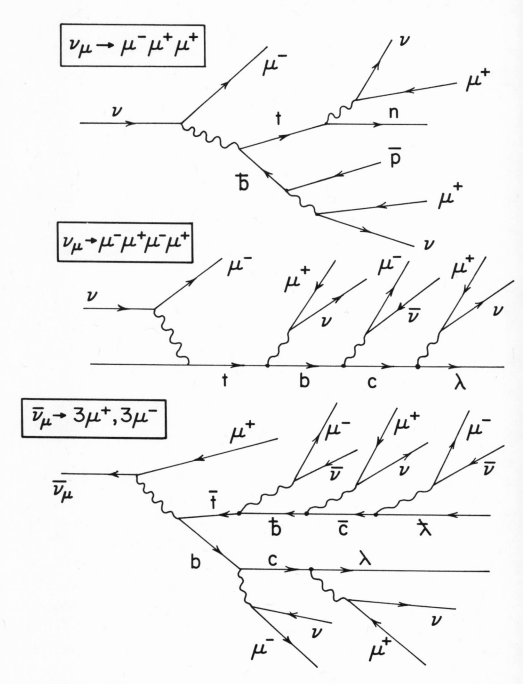

Figure 21   Weird multileptons in vector models.

$\bar{\nu} \rightarrow \mu^+\mu^+\mu^+\mu^-\mu^-\mu^-$. Some examples of tetra- and hexa-leptons (with $m_t > m_b > m_c$) are given in Fig. 21.

Whatever the correct theory of the weak inter-actions is, it seems safe to predict that equal sign dileptons are just the vanguard of a very interesting field of multi-prompt-lepton physics.

VI. NEUTRAL CURRENTS, THE HARD-FACT MODEL AND MADNESS

Neutral currents provide the simplest way to dis-tinguish between vector models and the standard model of Weinberg and Salam with purely left-handed charged currents. In strictly vectorlike models the neutral currents are by construction purely vector, while in the Weinberg-Salam model they are a V, A admixture. In both cases the neutral currents are diagonal in all quantum numbers and they do not share the energy de-pendence of the charged currents, caused by new hadron thresholds. The experimental situation is generally (and fairly) described as "in beautiful agreement with the Weinberg-Salam model." However, the conclusion of a careful analysis of the inclusive, single pion, and electron target data[42] is that vector theories are in equally good agreement with experiment. The crucial tests of vector theories are:

i) Parity violation should not be observed to $O(G)$ in the interaction between charged leptons and nucleons. Currently running experiments[44] aimed at the detection of parity-odd observables in atomic physics should fail.

ii) The differential cross sections for neutral current scattering of neutrinos and antineutrinos off a given target should coincide $d\sigma(\nu \rightarrow \nu) = d\sigma(\bar{\nu} \rightarrow \bar{\nu})$.

At this Conference, we have been presented pre-

liminary evidence[17] that the second above prediction
is violated: the energy distribution of recoiling
hadrons in ν and $\bar{\nu}$ induced neutral current events may
be different. On the other hand we have seen that the
evidence for a right-handed charged current in $\bar{\nu}$-induced
events is very compelling.

A phenomenological "hard-fact model" based on the
usual $SU(2) \otimes U(1)$ group would contain the following
doublets:

$$\begin{pmatrix} p \\ n_\theta \end{pmatrix}_L \quad \begin{pmatrix} c \\ \lambda_\theta \end{pmatrix}_L \; ; \begin{pmatrix} p \\ b \end{pmatrix}_R \quad . \tag{35}$$

In such a model the $\bar{p}p$ neutral current is purely
vector, while the $\bar{n}n$ current is a V, A admixture à la
Weinberg-Salam. To implement lepton-hadron symmetry
and cancel the anomalies an extra lepton doublet is
needed,

$$\begin{pmatrix} \nu_e \\ e \end{pmatrix}_L \quad \begin{pmatrix} \nu_\mu \\ \mu \end{pmatrix}_L \; ; \begin{pmatrix} L^0 \\ L^- \end{pmatrix}_R \quad . \tag{36}$$

Lo and behold, the extra charged lepton may be the
one allegedly discovered at SLAC.[45] There is even some
weak evidence that L^- decays may point towards a right-
handed coupling.[46] The hard-fact model as it stands does
the job that its name suggests.

If there is only one right handed coupling as in
(35) nothing can stop one from "Cabibbo-rotating" the p
quark,

$$\begin{pmatrix} p \cos \theta_R + m \sin \theta_R \\ b \end{pmatrix}_R \quad , \tag{37}$$

since there is no need for a right handed GIM mechanism.
The m quark can be the charmed quark and reduce its
strange decays, but m (for madness) is perhaps more
descriptive. In a model with the coupling (37) the
neutral currents are not diagonal, the cross section
slope $\sigma(\nu \rightarrow \nu)/E$ would be energy dependent and increase
as the threshold for madness is crossed. The increase
would be 3 times bigger for antineutrinos than
neutrinos. More serious and ambitious models that
incorporate the idea of madness can be constructed.
An example[47] is a model based on the simple group
SU(3), where the fractional charge assignment for
quarks and integer charge assignment for leptons are
natural. The quarks lie in triplets, among the left-
handed ones are:

$$(p, n_\theta, m)_L \ ; \ (c, \lambda_\theta, \tilde{m})_L \quad , \tag{38}$$

while leptons are in left-handed octets. This model
contains couplings of the type

$$\bar{m} \gamma_\mu (1+\gamma_5) n_\theta \bar{L}_0 \gamma^\mu (1+\gamma_5) \nu \quad , \tag{39}$$

that allow for the occurrence of neutral current nondiagonal process like

$$
\begin{array}{l}
\nu n \ \rightarrow \ m \ + \ L_0 \\
 \sqsupset \ \bar{L}_0 \ \nu \ + \ hadrons
\end{array}
\quad , \tag{40}
$$

where I have taken the extra lepton to be lighter than
the mad quark. In this particular model charged heavy
leptons also exist, whose semihadronic decays contribute
to the neutral currents and whose leptonic decays com-
plicate the study of charged currents. In general, the

slope of the neutrino neutral current cross section increases with energy, threshold effects for incident
antineutrinos are Cabibbo- or sea-suppressed.

REFERENCES AND FOOTNOTES

1. G. Snow, Nucl. Phys. B55, 191 (1973).

2. A. De Rújula and S. L. Glashow, Phys. Rev. D9, 180 (1973).

3. E. A. Paschos and V. A. Zacharov, Phys. Rev. D8, 215 (1973).

4. A. De Rújula, G. Georgi, S. L. Glashow and H. Quinn, Rev. Mod. Phys. 46, 391 (1974).

5. A. De Rújula and S. L. Glashow, Phys. Lett. 46B, 377 (1973).

6. M. A. Beg and A. Zee, Phys. Rev. Letters 30, 675 (1973), Phys. Rev. D8, 2334 (1973), Phys. Rev. D9, 1572 (1974). Riazzuddin and Fayyazuddin, Phys. Rev. D6, 2032 (1972). R. Budny and P. N. Sharback, Phys. Rev. D6, 3651 (1972). A. Love, D. V. Nanopolous and G. G. Ross, Nucl. Phys. B55, 33 (1973). S. Tamura and K. Fujii, Progr. Theor. Phys. 49, 995 (1973).

7. J. D. Bjorken and E. A. Paschos, Phys. Rev. 185, 1975 (1969) and D1, 3151 (1970).

8. H. D. Politzer, Phys. Rev. Letters 30, 1346 (1973). D. Gross and F. Wilczek, Phys. Rev. Letters 30, 1343 (1973). G. Parisi, Phys. Letters 43B, 207 (1973). D. Gross, Phys. Rev. Letters 32, 1071 (1974). A. De Rújula, H. Georgi and H. D. Politzer, Phys. Rev. D10, 2141 (1974).

9. Y. Watanabe et al., Phys. Rev. Letters 35, 898 (1975). C. Chang et al., Phys. Rev. Letters 35, 901 (1975).

10. Heavy quark masses also play an essential role in the analysis of e^+e^- annihilation in asymptotically free field theories. See A. De Rújula and H. Georgi, Phys. Rev., to be published, and E. Poggio, H. Quinn

and S. Weinberg, Phys. Rev., to be published.

11.   H. Georgi and H. D. Politzer, Harvard preprints.

12.   M. Barnett, Harvard preprints.

13.   S. L. Glashow, Nucl. Phys. 22, 579 (1961).
       S. Weinberg, Phys. Rev. Letters 19, 1264 (1967).
       A. Salam in *Elementary Particle Theory*, edited by
       N. Svartholm, Stockholm (1968).

14.   S. L. Glashow, J. Iliopoulos and L. Maiani, Phys.
       Rev. D2, 1285 (1970).

15.   R. Mohapatra, Phys. Rev. D6, 2023 (1972).
       A. De Rújula, H. Georgi and S. L. Glashow, Phys.
       Rev. Letters 35, 69 (1975) and Phys. Rev. D12, 3589
       (1975).
       G. Branco, T. Hagiwara and R. N. Mohapatra, CCNY-
       HEP7518 and COO-223B-84 (1975). These models have
       trouble with nonleptonic decays, see E. Golowich and
       B. R. Holstein, Phys. Rev. Letters 35, 831 (1975).
       A. Fernandez-Pacheco, A. Morales, R. Nũnez-Lagos and
       J. Sanchez-Guillén, GIFT report.

16.   H. Fritzch, M. Gell-Mann and P. Minkowski, Phys.
       Letters 59B, 256 (1975).
       F. Wilczek, A. Zee, R. L. Kingsley and S. B. Treiman,
       Phys. Rev., to be published. Vector models have also
       been constructed by S. Pakvasa, W. A. Simmons and S.
       F. Tuan, Phys. Rev. Letters 35, 702 (1975). Their
       version, however, is allegedly based on a model
       by H. Harari [Phys. Letters 57B, 265 (1975)] in
       which conventional SU(3) is also a symmetry of the
       heavy quarks. I do not know of a satisfactory way
       to incorporate this idea into a gauge theory of the
       strong interactions.

17.   F. Sciulli, these proceedings.

18.   M. Gell-Mann, these proceedings.

19.   A. Mann, these proceedings.

      A. Benvenutti *et al.*, HPWF76-1, to be published.

20.   H. D. Politzer, these proceedings.

21.   The heavy quark mass is defined as the position of
      the pole in the heavy quark propagator as computed
      to a given order in QCD perturbation theory when
      all renormalization subtractions are made at an
      Euclidean momentum. The mass parameter thus
      introduced finds its way into the expression for
      the cross section of real processes, where it
      describes the position and average shape of the
      threshold for production of particles containing
      the heavy quark.

22.   E. D. Bloom and F. J. Gilman, Phys. Rev. D4, 2901
      (1971).

23.   R. P. Feynman, unpublished.

      F. Gursey and S. Orfanidis, Nuovo Cimento 11A, 225
      (1972). O. W. Greenberg and D. Bhaumic, Phys.
      Rev. D4, 2048 (1971). G. Domokos, Phys. Rev. D4,
      3708 (1971).

      M. Briedenback and J. Kuti, unpublished.

      O. Nachtmann, Nucl. Phys. B63, 237 (1973).

      H. J. Schnitzer, Phys. Rev. D4, 1429 (1971).

24.   Based on unpublished calculations by M. Barnett.

25.   B. Aubert *et al.*, Phys. Rev. Letters 33, 984 (1974).

26.   T. Eichten *et al.*, Phys. Letters 46B, 274 (1973).

27.   Scaling and charge symmetry are enough to prove that
      the deviations of $\langle y \rangle$ from the naive value $1/4$ are
      $O(\alpha = \bar{Q}/Q)$. The result is:[2]

$$0.284 \pm 0.013 \sim \frac{1}{4} \left| 1 + \alpha \frac{3(3-R)}{8R} \right| \geq \langle y \rangle_{\nu} \geq \frac{1}{4} \left| 1 + \alpha \frac{(3-R)}{6R} \right| \sim 0.265 \pm 0.006.$$

28.   In asymptotically free field theories one expects the
      fractional amount α of antiquarks to increase with

$\langle Q^2 \rangle$ and thus with energy,[11] but the increase is far from enough to explain the y anomaly.

29. V. Barger, T. Weiler and R. J. N. Phillips, Phys. Rev. Letters 35, 692 (1975).
     Reasonable fits with the "slow" scaling variable of Ref. 4 have been obtained by S. Pakvasa, L. Pilachowski, W. A. Simmons and S. F. Tuan, Hawaii preprint.

30. $(p,b)_R$ is the "(1,0) fancy model" of Ref. 4, $(n,g)_R$ is the (0,1) model. A comparison of results from targets with different proton/neutron ratios provides a possible discrimination between the two models. In terms of ratios of cross sections the quantity $r = \sigma(\bar{\nu}p)/\sigma(\bar{\nu}(p+n))$ is predicted to have a low energy value $r = \eta/(1+\eta) \simeq 0.61$, where η, measure in electroproduction is $\eta = \dfrac{\int xp(x)dx}{\int xn(x)dx} \sim 1.56$. In the $(p,b)_R$ model r is energy independent, in the $(n,g)_R$ model r decreases to an asymptotic value of $r_\infty = (\eta+3)/4(\eta+1) \sim 0.45$. Tests in terms of $\langle y \rangle$ are more difficult.

31. If the new current is $(p,b)_R$ $\{(n,g)_R\}$, we can crudely expect the new resonances to be 1/4{4} times as large in the sense of $\int d\sigma(e^+e^- \to$ all), as ψ and ψ'.

32. C. Rubbia, Talk at the Palermo Conference.
     F. J. Scuilli, Talk at the APS Meeting in Seattle, 1975

33. FNAL-Michigan State Collaboration.
     Report of F. Roe at the Palmermo Conference.

34. A. Benvenutti et al., Phys. Rev. Letters 35, 1199, 1203, and 1249 (1975).

35. J. Von Krogh, report at the Wisconsin Meeting and private communication. The data is scanned for μ-e events. Six out of seven of the μ-e events

found have a K_S^0! (?).

36. Blietschau *et al.*, CERN preprint.

37. A. Pais and S. B. Treiman, Phys. Rev. Letters $\underline{35}$, 1206 (1975). The bound, based only on the assumption of a local S,P,T,V,A interaction in the hypothetical heavy lepton decay $L^0 \rightarrow \mu^+\mu^-\nu$, reads

$$0.47 \approx \frac{9-4\sqrt{2}}{7} \leq R \equiv \frac{\langle E(\mu^-)\rangle}{\langle E(\mu^+)\rangle} \leq \frac{9+4\sqrt{2}}{7} \approx 2.1.$$

The experimental value[34] is $R_{exp} = 3.7 \pm 0.65$, before subtraction of the antineutrino beam contamination. (After a plausible subtraction the result[34] is $R_{exp} = 6.1 \pm 0.8$).

38. L. M. Sehgal and P. M. Zerwas, Aachen preprint. J. LoSecco, Harvard preprint.

39. L. Ledermann, talk at the New York meeting of the APS (1976).

40. B. W. Lee, Report at the Conference on Gauge Theories and Modern Field Theory, Northeastern University, Boston, 1975.

41. T. Appelquist, A. De Rújula, S. L. Glashow and D. Politzer, Phys. Rev. Letters $\underline{34}$, 365 (1974). These authors predicted the existence of equal sign dileptons as a consequence of D^0-$\bar{D}^0$ mixing, which they incorrectly stated was a $\tan^2\theta_c$ effect in the standard model. Meanwhile, equal sign dimuons have been discovered, along with vector theories where D^0-$\bar{D}^0$ mixing can indeed be large.

42. A. Pais and S. B. Treiman, Phys. Rev. Letters, $\underline{35}$, 1556 (1975).

43. A. De Rújula *et al.*, in Ref. 15.

44. D. C. Soreide *et al.*, Phys. Rev. Letters $\underline{36}$, 352 (1976). M. A. Bouchiat and C. C. Bouchiat, Phys.

Letters 48B, 111 (1974), and J. Phys. (Paris) 35, 899 (1974) and 36, 493 (1975). P. G. H. Sandars in *Atomic Physics* 4, edited by G. zu Putlitz, E. W. Weber and A. Winnaker (Plenum, N.Y. 1975).

I. B. Khriplovich, JETP Letters 20, 315 (1974).

45. M. Perl *et al*., Phys. Rev. Letters 35, 1489 (1975).

46. S. Y. Park and A. Yildiz, to be published.

47. A. De Rújula, H. Georgi and S. L. Glashow, unpublished.

HOW TO COMPUTE THE CABIBBO ANGLE WITH SIX QUARKS*

S.L. Glashow

Lyman Laboratory of Physics

Harvard University, Cambridge, Mass. 02138

Once upon a time there were only three quarks.
Not so very long ago hadron spectroscopists thought
there were only three kinds (or flavors) of quarks, if
they believed in quarks at all. All known hadrons could
be built up out of u, d and s quarks. Each of the quark
flavors had the same strong interactions (now thought to
be mediated by color SU(3) gauge gluons) but differed in
mass and in weak and electromagnetic properties. But,
some of us were dissatisfied with this picture for the
following reasons:

The obvious lack of lepton-quark symmetry. There
are two weak doublets of left-handed leptons, while the
left-handed quarks formed one doublet and one singlet.
The existence of a fourth quark flavor with $Q = 2/3$
would allow the left-handed quarks to comprise two weak
doublets just like the leptons. The hadronic and
leptonic parts of the weak current would then look

---

*Work supported in part by the National Science
Foundation under Grant No. MPS75-20427.

much the same.

Certain weak decays of hadrons -- those violating
strangeness conservation and involving a pair of leptons with net charge zero -- are observed to be very
strongly suppressed. In old-fashioned weak-interaction
theory (before the renormalizable unification), these
processes occur only in second-order weak interactions,
but it nonetheless seems difficult to account for the
needed degree of suppression. In unified theories,
models with only three quark flavors are hopeless because the unwanted processes necessarily arise in order
G as well as order αG. With the introduction of a fourth
quark flavor, which is given the symmetric weak couplings
suggested by the lepton-quark analogy, a perfectly
natural origin is found for the absence of strangeness-
changing weak neutral currents.

Finally, there is the question of renormalizability
of the unified model of weak and electromagnetic interactions. A necessary condition for this is the absence
of triangle anomalies among the currents coupled to
gauge fields. This condition is not satisfied in a
model involving just the observed leptons and three
quark flavors. In many simple models, the condition
that there be no anomalies is just the statement that
the sum of the charges of the different quarks and
leptons is zero. Taking color into account, we see that
what is required is simply a fourth quark flavor with
$Q = 2/3$.

Lo: three distinct theoretical arguments suggest
the existence of the charmed quark, c. In any sensible
theory, the new quark must share the same strong interactions as the u, d, and s quarks. But where are the
hadrons containing one or more c quarks? Until the

November '74 revolution at BNL and SPEAR, no one had
claimed to have seen such a particle.

   Now we are four. Surely the simplest explanation
for the newly-discovered resonances is that they are
states of charmonium: $\bar{c}c$ bound states of a charmed
quark with its own anti-quark. The 3.1 GeV resonance
is the ground state of orthocharmonium; the 3.7 GeV
state is its first radial excitation. As predicted,
the ground state of paracharmonium exists with a mass
somewhat below the ground state of orthocharmonium:
the 2.8 GeV state. Also as predicted, p-wave states of
charmonium lie between the first two levels of orthocharmonium, and enjoy a variety of radiative decay modes.

   Given the mass of the charmonium states, we may
more-or-less reliably compute the masses of hadrons containing just one c or $\bar{c}$ quark. We predict the lightest
charmed mesons to lie near 1.9 GeV, and the lightest
charmed baryons to lie at 2.3 ± 0.2 GeV. This is consistent with the behavior of R in $e^{+}e^{-}$ annihilation: at
roughly 3.8 GeV there seems to be a threshold of the
appropriate nature. Moreover, several $J^{P} = 1^{-}$ states
lie above this threshold, and have widths hundreds of
times larger than the low-lying states. Surely; these
states are decaying into charmed hadrons; but experiment
has not yet demonstrated this fact.

   Why charm is like strangeness and why it is not.
Strange particles are produced in pairs (associated
production) by strong interactions, yet decay by weak
interactions: they exhibit the existence of a third
hadronic quantum number beyond u-number and d-number
(otherwise, quark charge and quark number): s-number or
strangeness. Strangeness is conserved by strong and
electromagnetic interactions, but weak interactions may

change strangeness by unity. *Da capo*, reading charm for strangeness and fourth for third.

The fundamental fermions form two isomorphic sets. On the one hand, there are the "relevant" particles: u and d quarks of three colors, electrons and their neutrinos. On the other, there is a similar set of particles of interest primarily to particle physicists: c and s quarks of three colors, muons and their neutrinos. Strangeness, charm, and muons are all of the same ilk. All interactions of the two sets are identical: relevant and irrelevant particles differ solely in mass. All phenomena may be explicable in terms of these sixteen particles, and the gauge fields of weak, strong, electromagnetic, and gravitational interactions (and, perhaps, some "Higgs" mesons).

But strange particles were found two decades ago, and charmed particles are just now being found. This is simply a reflection of the (unexplained and mysterious) fact that c quarks are much heavier than s quarks. Thus, charmed particles are heavier and harder to produce than strange particles, and they have awkwardly short lifetimes.

While the phenomenological distinctions between charmed particles and strange particles may be attributed to a mere accident of birth, the quark mass spectrum, there is also an historical and methodological distinction. Strangeness was invented to explain the systematics of an observed family of particles with strange properties. Charm was invented for purely theoretical reasons, rather distant from direct phenomenology. Charm predicts rather than just describes a new kind of matter.

Was charm discovered in neutrino physics? If

charm exists, single charmed hadrons must be produced
in the charged-current interactions of energetic
neutrinos and antineutrinos. The charmed hadron must
decay weakly, often yielding a strange final state.
The decay may yield an electron-neutrino pair, a muon-
neutrino pair, or may be non-leptonic. Evidence for
each of these decay modes has been reported.

At BNL, one fully reconstructed event may be
interpreted as the quasi-elastic production of a
charmed hadron (uuc) at about 2.4 GeV which decays non-
leptonically ($\Lambda \pi^+ \pi^+ \pi^+ \pi^-$).

In bubble chamers at CERN and at NAL, more than a
dozen neutrino-induced events have been seen involving
$\mu^- e^+$ and (sometimes) a strange particle. These are
easily interpreted as the production of a charmed
particle with mass $\sim$ 2 GeV with subsequent semi-electronic
decay.

Dimuon events involving the production of a pair
of oppositely charged muons have been reported by
several groups at NAL, produced both by neutrinos and
by antineutrinos. Kinematically, these events cannot
be explained in terms of the production of a new neutral
lepton. Again, they may be explained by the production
of charmed particles with masses $\sim$ 2 GeV which decay
semimuonically.

Finally, there is the high-y anomaly in the
charged-current antineutrino data at high energy, and
the failure of charge symmetry. Both of these *predicted* effects have been reported, and they may be
attributed to the $\bar{c}s$ part of the weak current and to
the consequent production of charmed hadrons.

The evidence for charm production in neutrino
interactions is compelling: Nature must make use of *at*

least four quark flavors. Although powerful theoretical
arguments pointed to the existence of the fourth
flavor, I know of no argument that limits Nature to just
four.

That that, that that that refers to, is not the
that to which I refer; or; Is more than three infinity?
When it was becoming clear that the J or ψ was made up
of a new kind of quark, Phil Morrison remarked to me
that while he could have swallowed three quark flavors,
four was simply absurd. To be elegant, there had to be
an infinite number of flavors. Now, if the requirement
of (permanent) asymptotic freedom makes sense, we cannot
go quite so far; we are limited to no more than fifteen
flavors. A *priori,* any number of flavors between four
and fifteen is admissible. *"How many quark flavors?* is
the most immediate problem for the hadron spectroscopist.
The new flavor (or flavors) responsible for J and ψ may
not be the same as the new flavor involved in the GIM
cancellation, and more than one new hadronic quantum
number may well show up in neutrino interactions.

At this point, an embarrassing theoretical weakness
is revealed. Not only do we not know how many flavors
nature uses, but we have no real understanding of flavor
at all. The ultimate (for the moment) spectroscopy,
that of quarks and leptons, remains a complete mystery.
It will only be a helpful first step to know whether
there are only four quark flavors, or fourteen. We have
lived with the known existence of two flavors (hence,
protons and neutrons) since the year I was born; and
a theory of flavor still does not seem to be in the
offing.

While four quarks seem to suffice for a qualitative
explanation of all the recent exciting developments,

there are straws in the wind suggesting that more than
four flavors may be needed.

Lepton-quark analogies, revisited.   Martin Perl,
so it is said, has discovered a third charged lepton.
His new lepton decays weakly, and thus is involved in
weak as well as electromagnetic interactions.   Presumably, there are thus *three* weak doublets of leptons,
and so there should be three doublets of quark flavors
making up *six* quark flavors.   Arguments concerning
triangle anomalies suggest the same result.   (Less
elegantly but more perversely, we could make do without
a third neutrino by forcing the new lepton to share the
electron and muon neutrinos.   Then, lepton-quark analogies would suggest only *five* quark flavors.)   Should
even more charged leptons be found, we should expect even
more quark flavors.

Vector-like theories.   In a vector-like theory, the
Lagrangian is invariant under space reflection, and observed parity violation is due to spontaneous symmetry
breaking.   Phenomenologically correct vector-like
theories cannot be made with just four quark flavors.
Many groups (Harvard, Princeton, CUNY, CalTech, Hawaii,
and others) recently have suggested vector-like models
involving six: u, d, s and c and two new flavors t and
b for truth and beauty.   (Flavors u, c, and t have
$Q = 2/3$, while d, s and b have $Q = -1/3$.)

These models, in which all six left-handed flavors
and all six right-handed flavors belong to weak doublets, make a number of immediate and specific predictions.
There must be no parity-violating interaction between
electrons and hadrons to order G.   Thus, no parity-
violation should be found in atomic physics experiments
such as those of Bouchiat and Bouchiat and of Sandars.

Similarly, there should be no parity-violating effects in e^+e^- annihilation due to Z^o mediation.

In these models, the neutral-current couplings of neutrinos to hadrons should be pure V and neutrino and antineutrino neutral current cross sections should be identical.

Another kind of vector-like theory can be constructed in which some of the quarks transform as weak singlets. Phenomenological models of M. Barnett, and theories based on the exceptional Lie algebra E(m) due to Gürsey and Ramond are of this kind. Besides u, d, s, and c, these models involve two additional flavors with Q = -1/3.

Is fancy found? In the standard model, just four quarks with just left-handed currents, neutrino or antineutrino interactions off valence quarks make little contribution to the production of hadrons with new quantum numbers. Neutrino effects are suppressed by $\tan^2\theta$, and antineutrino effects are absent. The observed high-energy effects must be due to sea quarks. In other models, it is possible to couple right-handed u and d quarks to new quarks, charmed or otherwise, without the necessity for any Cabibbo-like suppression. Generically, De Rujula, Georgi, Quinn and I christened such couplings "fancy". In fanciful models, including the popular six-quark vector-like models, valence quarks do contribute to the production of new hadrons. As we warned, the effects of fancy may easily be confused with the effects of charm.

Some experimental data seem to indicate more than four quarks. The high-y anomaly in antineutrino data is an indication that hadrons bearing a new quantum number are being produced: perhaps fancy, perhaps charm.

The analysis presented at this meeting by De Rujula
(based on works by Georgi and Politzer, and by Barnett)
indicates that the effect reported is simply too large
to be due to charm, but is precisely the predicted
effect of fancy. At least five quark flavors seem to
be needed.

The value of R in e^+e^- annihilation, suitably
smeared by Poggio, Quinn and Weinberg, and by Georgi
and De Rujula, also provides a means of counting quark
flavors. Once again, five or six flavors seem necessary
to explain the data.

Finally, there are the "wrong sign dimuons" reported by the Harvard-Penn-Wisconsin collaboration.
Though possibly explained as associated production of
charmed hadrons by conventional neutrino couplings, it
seems to me that a more likely explanation can be found
in fancy six-or-more quark models where cascades of
weak decays may take place, or where significant $D^0\bar{D}^0$
mixing may be induced.

Reasonable men could agree that there are only
three viable possibilities for models based on six or
fewer quarks:

The Standard Model. A fancy-free model with four
flavors (u,d,s,c) and left-handed weak currents. The
only experimental conflict this model encounters is the
large cross section for new hadron production --
suggestive of production off valence quarks, and hence,
of fancy -- that may be deduced from the large observed
high-y anomaly in antineutrino interactions. (See contributions of A. Mann and A. De Rujula to this Conference.)
This model may be extended to six quarks to provide
lepton-quark symmetry (see above), or to encompass CP-
violation (see below).

The Vector Model. A fancy model with six flavors
(u,d,s,c,t,b) grouped into three left-handed doublets
and three right-handed doublets. This model encounters
difficulty with experiment only in the allegation by F.
Sciulli (at this conference) that certain data seem not
to confirm that the hadronic neutral current is purely
vector.

The Expedient Model needs no more than five quarks
to fit all current data. The left-handed current is
that of the standard model, while the right-handed
current involves the single doublet (u,b). Evidently,
"beauty" is produced off valence u quarks by antineutrinos but not by neutrinos: the HPW data on y-distributions and dimuons may be fit. On the other hand, the
hadronic neutral current is *not* purely vector. The expedient model, with $m_c \sim m_b$, was originally proposed by
M. Barnett; however, to fit the data now available we
need $m_b \geqslant 2m_c \sim 3$ GeV. An extended vector-like expedient model involving the additional right-handed
doublet (c,b') and yet one more new charged lepton was
also first proposed by Barnett, and independently arrived
at by Gürsey and Ramond. Both new leptons must have unconventional V+A couplings to their neutrinos, and hence
should display vanishing Michel parameters in their
leptonic decays.

If all of the neutrino data described at this
conference are accepted at face value, then it is only
the expedient model that remains. On the other hand,
e^+e^- annihilation data show no evidence for a narrow
resonance at $\geqslant 6$ GeV corresponding to the $\bar{b}b$ bound state.
Something has got to change.

Who violates CP? In the standard model, there is
no room for CP violation in the coupling of the

intermediate vector bosons. CP violation *must* be
mediated by the Higgs bosons. I find this inelegant
and unsatisfying, although I am sure there are others
who feel just the opposite. I would prefer that CP
violation, like Cabibbo mixing, should be attributable
to the form of the quark mass matrix. (Depending, if
you will, on the vacuum expectation values of the Higgs
fields, but not directly on their dynamics.)

First, suppose that there are *no* right-handed
currents. Then, with just four quark flavors, there
can be no CP violation intrinsic to the weak current.
However, with six quark flavors put into three left-
handed doublets, there is room for CP violation. This
was recently discussed by Maiani and by Pakvasa and
Sugawara. Indeed, five quark flavors (u,d,s,c and
either t or b) suffice to construct such a model. A
remarkable prediction of these models is that the
nucleon's electric dipole moment only appears at order
G^2, and is too small to be detected even by Norman
Ramsey.

In a vector-like theory with six quarks, it is of
course very easy to incorporate CP violation. The most
general charged current then involves thirteen convention-independent "Cabibbo angles", of which seven
are intrinsically CP violating.

What is calculable? "Unification" has a nice
sound, yet we are being led to models with more and more
parameters. It seems reasonable to believe that they
are not all independent, and that relations will be
found among them. In order to know what kinds of relations to expect, it is necessary to know more than we
do. At this point, all we can do is make guesses. Let
me make the following guess. To zeroth order in α,

there is no flavor mixing in the weak current: the
members of left-handed (and right-handed) weak doublets
can be chosen as *pure* flavors. Effects like Cabibbo
mixing are assumed to be calculable order α effects.
With this assumption, we may explain:

*How to calculate the Cabibbo angle with six (or
more) quark flavors.* First, we must decide what are
the unmixed zeroth-order weak doublets. With six
quarks, and with the assumption that the induced flavor
mixing is a small effect, the three left-handed doublets
are uniquely determined: (u,d); (c,s); and (t,b). There
are just two phenomenologically acceptable assignments
to right-handed doublets: (u,b); (c,d); and (t,s) *or*
((u,b); (c,s); and (t,d). We adopt the former choice.
Clearly, the latter choice leaves c-number plus s-number
conserved to all orders in α, a phenomenologically un-
acceptable result.

Although we assume unmixed flavors to order α^{o}, we
allow the possibility of CP-violating phases. It is not
so evident but true that a most general weak current
satisfying the SU(2) algebra with these multiplet assignments is:

$$\bar{u}\gamma_L d + \bar{c}\gamma_L s + \bar{t}\gamma_L b + e^{i\chi}\bar{u}\gamma_R b + \bar{c}\gamma_R d + \bar{t}\gamma_R s \quad ,$$

where $\gamma_{L,R} = \frac{1}{2}\gamma_\mu(1 \pm \gamma_5)$. There is just one CP-violating
convention-independent phase. In an extension of this
model to more than three doublets of quarks, the assignments into right-handed doublets is a permutation of
the assignments into left-handed doublets. The number
of independent CP-violating phases is the number of
cycles of this permutation, in the present model, the
permutation is cyclic.

To order α, there will be diagrams (involving emission and resorption of a charged W) contributing to the quark mass operator, thus leading to flavor mixing and in particular to the appearance of the Cabibbo angle.

These diagrams diverge. One hopes that in a richer, more realistic model, these infinities may be avoided and that the Cabibbo angle will turn out to be finite and calculable. If weak $SU(2)$ is supplanted by weak chiral $SU(2)$, the order α diagrams can naturally be arranged to converge, yet infinities reappear in the two-loop approximation. Models based on the weak gauge algebra 3 $SU(2)$ + $U(1)$, although inelegant, can yield finite results. But, these models (it seems) must involve at least eight quark flavors. One such scheme has been devised by H. Georgi.

We put these technical difficulties aside, and return to an estimate of the flavor mixing in the six quark model produced in order α. We replace the dimensionless divergent integral by an unknown parameter λ. The corrections to the weak current are

$$A\bar{u}\gamma_L s + B\bar{u}\gamma_R s + C\bar{u}\gamma_R d + \ldots \, ,$$

where the omitted terms do not contribute to decays of ordinary hadrons involving only u, d and s quarks. We find

$$A = \alpha\lambda m_c m_d [(m_s^2 - m_d^2)^{-1} - (m_c^2 - m_u^2)^{-1}] \quad ,$$

$$B = \alpha\lambda e^{i\chi} m_b m_t [(m_t^2 - m_u^2)^{-1} + (m_b^2 - m_s^2)^{-1}] \quad ,$$

$$C = \alpha\lambda m_d m_u [(m_c^2 - m_u^2)^{-1} + (m_b^2 - m_d^2)^{-1}] \quad ,$$

with plausible values for the quark masses, we obtain
the correct value for the left-handed Cabibbo angle
with $\lambda \approx 6$.

However, it is then necessarily so that B is not
small: we find $|B| \sim 1/2$ A. Thus, there is a significant
right-handed contribution to the strangeness-violating
current. To be compatible with the experimental fact
that $\theta_V \approx \theta_A$ we must require that the CP violation is
maximal, and put $\eta = \pi/2$. The only direct experimental
consequence of this is to predict large CP violation in
the semi-leptonic decay of strange hadrons. *We predict
a large relative phase between* G_V *and* G_A *in hyperon beta
decay, of order* $45°$. There is little experimental data
with which to test this result.

Finally, we note that there is a small CP-conserving
but right-handed contribution to the strangeness-
conserving current. This will lead to a small ($\approx 2\%$)
change in the ratio G_V/G_A, which is entirely masked by
conventional strong interaction effects. (The Adler-
Weissberger relation is simply not reliable enough to
demonstrate the existence of this effect.)

Warning. This calculation is premature. We do
not know if we are working with the correct quark
flavors. We do not know if the Cabibbo angle is a
calculable order α effect, and we certainly do not have
a plausible theory in which it is. All we intend is a
demonstration of how some things might work out.

HADRON STATES IN THE M.I.T. BAG MODEL[*]

K. Johnson

Massachusetts Institute of Technology

Cambridge, Massachusetts   02139

The M.I.T. Bag Model[1] is a model of permanently confined hadron constituents. The constituent confinement is simple and can be understood classically. It does not hinge upon special quantum numbers. The confinement of quark quantum numbers is also easily accommodated but requires a special model, chromodynamics, that is, quarks coupled by a non-Abelian vector gauge field. However the confinement of quark quantum numbers can also be easily understood classically in this case.

Because our model is not based on a conventional local field theory, many regard it as necessarily phenomenological. That is, they hope that conventional local field theory will be capable of describing hadrons as composed of confined quarks. However, at present this hoped for result from local field theory is still but a dream.

————————

[*]This work is supported in part through funds provided by ERDA under Contract AT(11-1)-3069.

Our attitude about our model is that it may be a
phenomenological model, that is a crude approximate way
of determining the consequences of a more accurate theory,
or it may be an accurate theory (that is, less crude).

It would be hopelessly naive to assume that any
model has any ultimate truth, so all are phenomenological
on some scale or other. In any case, I believe that it
is rather extreme to adopt the position that the quantization of a local classical field theory is the only way
that an accurate quantitative description of strong interactions could be obtained.

Today I would like to briefly review our main assumptions, and then turn immediately to a summary of the
principal consequences for hadron spectroscopy. Because
of the success of quark-parton phenomenology, we assume
that the quark amplitudes look as in Fig. 1 within hadrons
that is, the quarks move more or less freely inside and ar
small outside. Such a wave function would be a consequenc
of a very high confining potential exterior to the region
occupied by the particle. The particle is therefore extended in space. Since a picture like this implies that
the quark is not found in the region where the forces are
very strong, the main contribution to the mass of the
hadron will come from the quark amplitude on the inside
where the quarks move freely. Our model is based on the
assumption that in the classical limit this result is
exact: we replace the strong force by a local, covariant,
boundary condition which is consistent with the requirement that the quark amplitude be zero outside of the
hadron. We assume classically that the surface is infinitely sharp. It is classically sharp since we wish
the model to be one consistent with relativity and
causality. (That is also why elementary particles are

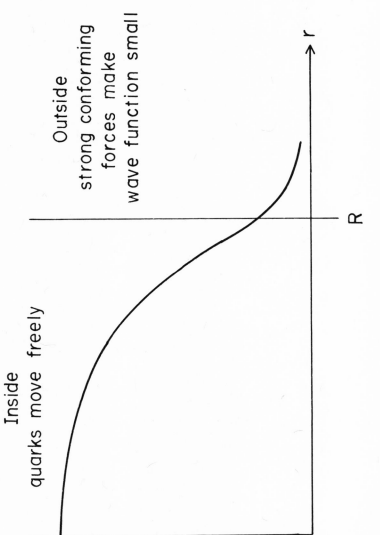

Figure 1   Schematic Ground State Quark Wave Function in Hadron

assumed to be classical points in the ordinary theory).

As a consequence of the surface boundary condition, the quarks exert a pressure on the confining surface. If n_μ is the space-time normal to the surface (n_μ = space normal in system where the surface point is at rest), it can be shown[1] that at a surface point,

$$n_\mu \, T^{\mu\nu}(x) = n^\nu P(x) \ ,$$

(1)

where $T^{\mu\nu}(x)$ is the local stress tensor of the hadronic constituent fields and $P(x)$ is a local, invariant function of the same fields. Physically, $P(x)$ is a pressure. The fields move inside of the hadrons so that $\partial_\mu T^{\mu\nu}(x) = 0$ as in conventional field theory. In order to conserve energy and momentum we take the total energy and momentum of the hadron to be given by the expression,

$$T^{\mu\nu}_{Hadron} = T^{\mu\nu} - g^{\mu\nu} B \ , \ \text{inside}$$

$$= \ 0 \qquad \qquad , \text{outside}$$

(2)

where $T^{\mu\nu}$ is the stress tensor of the constituents and $-g^{\mu\nu}B$ is a constant tensor. Therefore $\partial_\mu(g^{\mu\nu}B) = 0$ trivially. We choose B so that no flow of momentum or energy comes through the surface, that is

$$n_\mu \, T^{\mu\nu}_{Hadron} = 0 = n_\mu \, T^{\mu\nu} - n^\nu B$$

(3)

on the surface. Therefore by (1),

$$0 = n^\nu(P(x)-B),$$

(4)

the equation $P(x) = B$ is a local covariant contour map

for the surface of the particle. Since it is local, the
theory is classically consistent with causality. Thus,
easy to show that if B>0, the surface remains space-like,
that is, all parts move with speed less than the speed of
light. One can think of B as a confining pressure. B
is a constant with dimensions of pressure or energy/
volume which sets the scale for hadronic physics. We
call it the universal bag constant. In units where
$\hbar = c = 1, B^{\frac{1}{4}}$ is a mass (~145MeV). At this point many ask for
the "origin" of B. No model can explain the "origin" of
the parameters which are contained within it. The origin
of B must lie in physics at a deeper level. As already
noted, we have an open mind about how deep one must go
to find the origin of B.

If we allow the hadronic constituents to be colored
quarks and colored gluons coupled in the standard way,
it is a simple consequence of the chromatic analogue of
Gauss's law that the total color of a hadron must be
zero. Hence the strong interaction (fission of bags) can
take place only if the fission occurs into color singlets.
Hence there are no hadrons with quark quantum numbers.
Thus, quark confinement is an automatic consequence of
the finite spacial extension of hadronic matter in the
context of a model with colored constituents.

Let us now summarize some of the immediate consequences of our model. If we assume that the wave function
for the hadron peaks about some spherical radius R, we
may obtain an approximate version of our system by simply
quantizing the quark gluon fields within a spherical
cavity with this radius.

In this case the energy of the constituents will be
$E(R)$. According to (2), the total energy will be

$$E = E(R) + \frac{4\pi}{3} BR^3 \quad . \tag{5}$$

Since the pressure of the constituents is $-\frac{\partial E(R)}{\partial V}$ we find that R is given by minimizing (5),

$$-\frac{\partial E(R)}{\partial R} - 4\pi BR^2 = 0 \quad . \tag{6}$$

we thus obtain a radius for each particle.

We have calculated the masses of the classic SU(6) ground states of mesons and baryons in the approximation for the particle wave function described above, also taking into account the effect of color gluon exchange between quarks in lowest order.[2] When the gluon exchange is omitted the states form degenerate SU(6) multiplets. Since the gluons are vector fields, the exchange produces the chromatic analogue of the hyperfine interaction of ordinary electrodynamics. We fit the gluon coupling constant to the SU(6) breaking, in particular, we choose to fit the $\Delta(1280)$, proton mass splitting.

In addition to the pressure produced by the energy of the valence quarks, we have also added a term to the energy which is associated with the "sea" or vacuum of quark-antiquark pairs and gluons. Since most of those are massless fields, we expect the dominant sea contribution to be dimensionally of the form $-\frac{z}{R}$, where z is a dimensionless constant. We have treated z as a phenomenological constant fit without a priori restriction to the sign. We find, z>0 and ~1.8. This should be compared to the energy of a single massless valence quark where $E_{valence} = \frac{2.04}{R}$.

Although z can in principle be calculated theoretically, at present no computation has been done. A crude

estimate has been made in the colored quark gluon model
which is consistent with the sign and magnitude of z
found in the phenomenological analysis.

The final parameter fit was the strange quark mass.
We fixed it to obtain the observed separation between the
Ω^- and proton, the largest SU(3) non-conserving separation
within the baryon super multiplet. We find $m_s \sim 280$ MEV
(with $m_u = m_d = 0$).

The results of our calculations[2] are presented in
Fig. 2. We see that the calculated masses are in quite
good agreement with ordering of the observed hadron levels
including subtle effects such as $m_\Sigma > m_\Lambda$. The state which
is quantitatively the worst is naturally the π. We made
no attempt to make this state have a low mass. It comes
out naturally to be very light because the color hyperfine
interaction is stronger in $q\bar{q}$ states than in qqq states,
and is proportional to $\frac{1}{R}$ which is larger in low mass
states since the field pressure is lower thus making the
hadron smaller. However, we also would expect that the
static cavity approximation makes the best sense when R
is large in comparison to the size related to the Compton
wave length of the particle. For the π, the low mass
makes $\frac{1}{m}$ very large. Hence we do not believe it is
sensible to fiddle with parameters to attempt to obtain m_π
more accurately, at least in the context of the static
cavity approximation.

In the above cavity calculations we have omitted
reference to the flavorless pseudoscalar states (η, η').
If we calculate their masses in the same way as for the
flavored pseudoscalars we would find $\eta = (u\bar{u} + D\bar{D})\frac{1}{\sqrt{2}}$ and
with a mass equal to the π, and $\eta' = s\bar{s}$ and with a mass
a bit larger than the K. It is a well known problem within
the quark model to obtain an adequate description for the

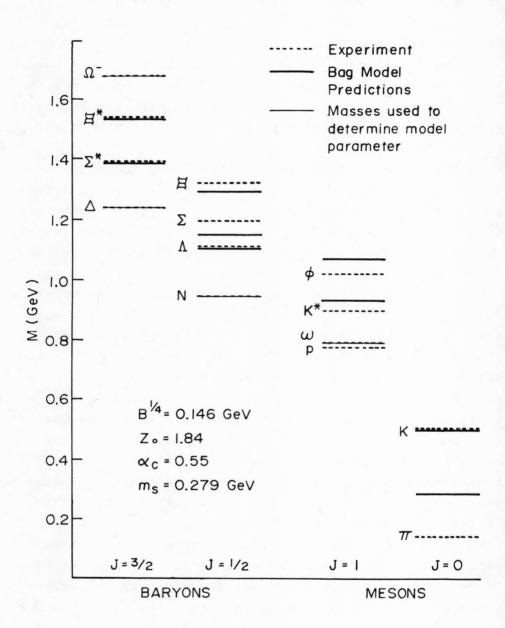

Figure 2

0^- flavorless states, and our model shares this difficulty.
We believe (the same belief is also held by others) that
the resolution of this problem must be associated with
the mixing of the SU(3) flavor singlet component of these
states with pure gluon states. In the cavity approximation, the lowest 0^- gluon, color singlet state, contains
two gluons, one with energy $\frac{2.74}{R}$ (TE mode) and one with
energy $(\frac{4.49}{R})$ (TM). Although the 0th order field energy
of these states is $(\frac{7.23}{R})$ in comparison to the 0th
order quark antiquark energy of $\frac{2\times2.04=4.08}{R}$, they are
close enough and the coupling is strong enough that
degenerate perturbation theory should be applied. Work
on this is in progress.

To see how our parameters relate to a completely
different regime in hadron phenomenology,[3] we have also
studied the properties of Regge trajectories. For large
J the trajectories are linear, with a slope given by

$$\alpha^1 = \frac{1}{8\sqrt{2\pi}^{3/2}} \frac{1}{\sqrt{C}} \frac{1}{\sqrt{\alpha_c}} \frac{1}{\sqrt{B}} . \qquad (7)$$

The particles on the trajectories correspond to spinning,
approximately linear bags with opposite colors on the ends.
The Casimir operator corresponding to the color representation on one end is $C = \sum_a \Lambda_a^2$. The meson and baryon representations have the largest slope, in both cases there
is a quark on one end. In the case of a meson there is
an antiquark (color $\bar{3}$) on the other. In the case of a
baryon there is a diquark with the same color as the antiquark on the opposite end. The slopes for meson and
baryon trajectories are therefore the same, $C_{\bar{3}} = C_3 =$
$\sum_a (\frac{\lambda_a}{2})^2 = \frac{4}{3}$. If we take in this case $B^{\frac{1}{4}}$ and α_c as determined
from the ground state spectroscopy, we find $\alpha' = .9(\text{GeV})^{-2}$

for the meson and baryon trajectories which is consistent
with the observed trajectories which are linear even for
low J. We may note that if we use units where $\hbar \neq 0$, the
expression (7) for α' makes no reference to $\hbar$. Therefore
if the approximate linearity of the trajectory is also
true for low J (which we have not shown), then it would
not be surprising that the above semi-classical calculation yields the observed slope.

   If we assume that the true trajectories are linear
for all J, then we can obtain a single estimate for the
intercepts from the cavity approximation calculations
discussed earlier. One of the most interesting of these
estimates concerns the trajectory with color octets on
the ends, which because $C_8 = 3$, has a slope $\alpha_8' = \frac{2}{3}\alpha_3' =$
$.6(GeV)^{-2}$. This trajectory is obtained with a colored
gluon on each end and may be expected to be the one on
which the Pomeron lies. If we calculate M^2 for the 2^+
gluon state in the cavity approximation, and use the
formula

$$J = \alpha'M^2 + \alpha_o, \qquad (8)$$

with $J = 2$ to evaluate α_o, we find $\alpha_o = 1.01.$[3]

   Finally, armed with the parameters provided by
the calculations of these classic states we have begun
to investigate unconventional quark model states. Unfortunately, time will not permit me to discuss our
results here[4].

REFERENCES

1.   A. Chodos, R. L. Jaffe, K. Johnson, C. B. Thorn,
     V. F. Weisskopf, Phys. Rev. D9, 3471 (1974).

2.   T. DeGrand, R. L. Jaffe, K. Johnson, J. Kiskis,
     Phys. Rev. D12, 2060 (1975).

3.   K. Johnson, C. B. Thorn, Phys. Rev. D, to be publish-
     ed.

4.   R. L. Jaffe, K. Johnson, Phys. Lett. 60B, 201 (1976)
     and to be published.

Murray Gell-Mann's Joint Yogurt Project with Hoja is being inaugurated.

Dr. Minkowski is witnessing it.

RECENT RESULTS FROM SPEAR*

W. Tanenbaum

Stanford Linear Accelerator Center

Stanford University, Stanford, California   94305

I will discuss primarily new e^+e^- colliding beam results from the SPEAR Magnetic Detector[1,2] since the Lepton-Photon Symposium last August.[3,4,5] Due to time limitations, I will not cover certain areas where there are essentially no new results, such as the anomalous μ-e events.[6] The topics I will discuss are:

I.   Total cross section in the 4 GeV region

II.   Exclusive multipion channels

III. ψ' decays

IV.   Upper limits on high mass resonances

This talk is not meant to be a complete review of the experimental situation. For such a review, see References 3, 4, and 5.

I. TOTAL CROSS SECTION IN THE 4 GeV REGION

Figure 1 shows the ratio

$$R = \sigma_{e^+e^- \to hadrons} / \sigma_{e^+e^- \mu^+\mu^-} \text{ vs } E_{c.m.} \quad ,$$

*Work supported by U.S. Energy Research and Development Administration.

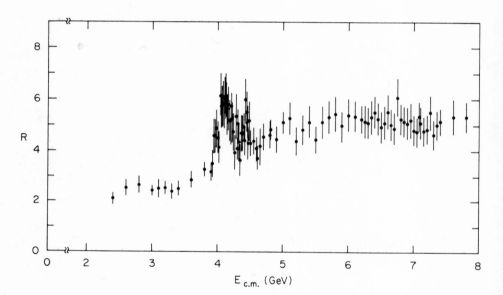

Figure 1   R versus E$_{c.m.}$ in coarse steps.

as was presented last August. The data have been
corrected to remove the radiative tails of the narrow
ψ and ψ' resonances and final states arising from two-
photon processes. The error bars include statistical
errors and ± 10% point-to-point systematics. There is
also an overall ± 10% normalization uncertainty.

We see a rich structure in the 4 GeV region. We
have recently taken more data. Figure 2 shows the
ratio R in the 4 GeV region. The errors shown are
statistical only. The open points represent new data.
We see a complex and as yet unexplained structure in
the region between 3.9 and 4.1 GeV. The data in this
region are well fitted by several different assumptions
about the masses and widths of resonances present. All
that can be said at this point is that more data are
needed to unambiguously resolve the structure in this
region.

In the region around 4.4 GeV, the existence of
a resonance, first reported at the Symposium, has been
confirmed by new data. Figure 3 shows the ratio R from
4.3 through 4.5 GeV.

The data points have been corrected for radiative
effects except those involving the 4.4 resonance itself.
Radiative corrections involving the resonance itself
are not used in correcting the data points; rather they
are included by correcting the expected Breit-Wigner
distribution. The solid curve in Fig. 3 indicates the
expected shape of a single radiatively corrected Breit-
Wigner fitted to the data. The data are well accounted
for by a single resonance. The parameters of the resonance are shown in Table I. For comparison purposes,
the corresponding parameters of the ψ and ψ' resonances
are also shown.

TABLE I

Resonance Parameters

	4.4	ψ	ψ'
Mass	4414 ± 7 MeV	3095 ± 4 MeV	3684 ± 5 MeV
Γ	33 ± 10 MeV	69 ± 15 keV	225 ± 56 keV
Γ_{ee}	440 ± 140 eV	4.8 ± 0.6 keV	2.2 ± 0.3 keV
$\int \sigma_H(E)dE$	≈ 500 nb · MeV	10400 nb · MeV	3700 nb · MeV

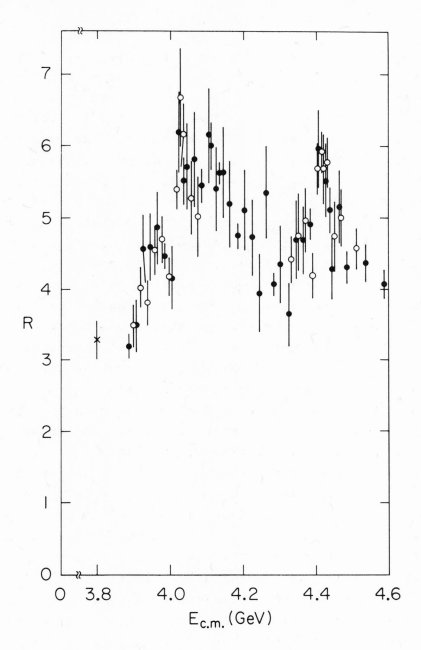

Figure 2   R versus $E_{c.m.}$ in the 4 GeV region.   The open
circles are new data.

II. EXCLUSIVE MULTIPION CHANNELS

In Fig. 1 we see not only the structure in the 4
GeV region but also that the ratio R has a relatively
constant value of 5 above the 4 GeV region, while below
the 4 GeV region R has a value of about 2.5. This
step rise in R suggests that new hadronic degrees of
freedom are opening in the 4 GeV region. It is of extreme importance to discover what, if anything, dis-
tinguishes the "new physics" from the "old physics."

I will review briefly a little of what we already
know about this question. All of this has been covered
previously by R. Schwitters in Ref. 3. The mean charged
multiplicity versus energy is plotted in Fig. 4. The
values shown have been corrected for loss of charged
particles due to geometric acceptance and trigger
biases. The data are consistent with a logarithmic
rise with c.m. energy. There is no evidence for a change
in this behavior at 4 GeV although the level of uncertainty is large. Figure 5 shows the mean energy of th
observed tracks. Only tracks with 3 or more prongs
were considered and all tracks were assumed to be pions.
The mean charged track energy increases with c.m. energy except right near a beam energy of 4 GeV, where
there is evidence for a leveling off. This may indicate
a small but sudden increase in the total (charged +
neutral) multiplicity. The mean fraction of total
energy appearing as charged particles is plotted in
Fig. 6. Three or more prong events were used, and pion
masses were assigned to all tracks. Again, the data
have been corrected for loss of charged particles. The
fraction of charged energy falls with increasing c.m.
energy. There is no strong evidence for a discontinuity

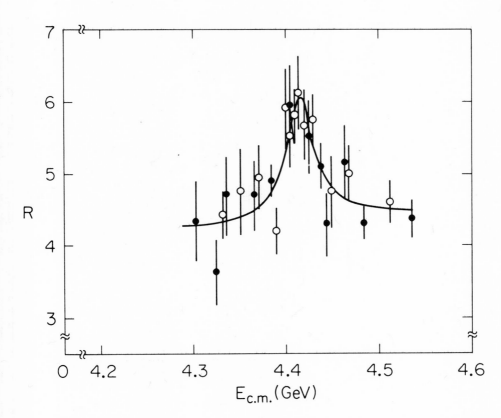

Figure 3   R versus $E_{c.m.}$ near 4.4 GeV.   The solid curve is
a Breit-Wigner fitted to the data.

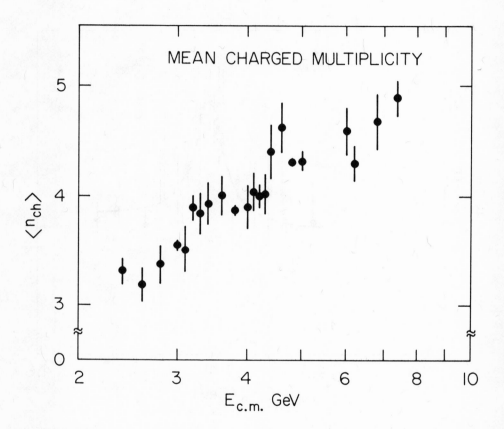

Figure 4   Mean charged multiplicity $\langle n_{ch} \rangle$ vs $E_{c.m.}$.

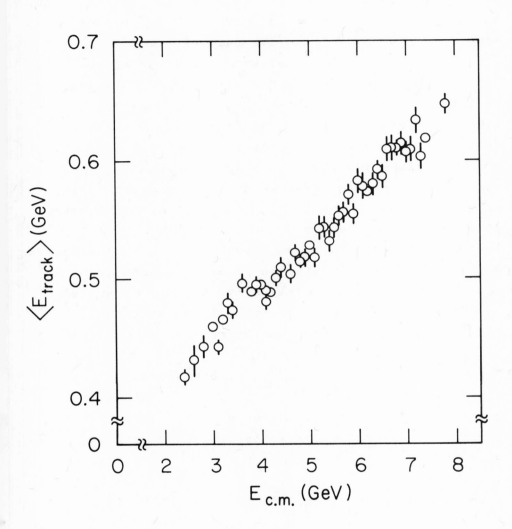

Figure 5   Mean energy per track for $\geq$ 3 prong events vs $E_{c.m.}$.

in the 4 GeV region. In short, there seems to be a hint
that the neutral multiplicity increases suddenly in the
4 GeV region, but the evidence is not conclusive. I do
not have sufficient time to talk about the inclusive
momentum spectra or inclusive particle production.

We can attempt to learn more by studying exclusive
final state production as a function of c.m. energy.
We thus hope to learn whether the "new physics"
favors different final states from the old physics.
Unfortunately, apart from the ψ and ψ' resonances,
cross sections are low. The only final states that we
can study with meaningful accuracy over a broad range of
energies are multipion final states with all charged
pions, specifically $2(\pi^+\pi^-)$ and $3(\pi^+\pi^-)$. States with
one missing π^o can also be reconstructed, but there are
experimental difficulties in separating them from background. We have previously shown that the ψ and ψ'
resonances do not decay directly to an even number of
pions. Thus, data taken at the resonances can be
included in our analysis below.

Fig. 7a shows the exclusive cross section for
$e^+e^- \rightarrow 2(\pi^+\pi^-)$ as a function of $E_{c.m.}$. The data are
consistent with a smooth exponential falloff ($\sigma \propto s^{-2.8\pm0.5}$)
There is no evidence for a discontinuity or kink in the
4 GeV region. Fig. 7b shows the exclusive cross section
for $e^+e^- \rightarrow 3(\pi^+\pi^-)$. Again the data are consistent with
a smooth exponential falloff ($\sigma \propto s^{-2.3\pm0.8}$). If
"new physics" liked to decay into $2(\pi^+\pi^-)$ or $3(\pi^+\pi^-)$ as
much as old physics, we might expect a factor of two
rise in the exclusive cross sections around 4 GeV where
the ratio R rises by a factor of two. Although not conclusive, the data tend to show the "new physics" does not
like to yield all pion final states with an even number

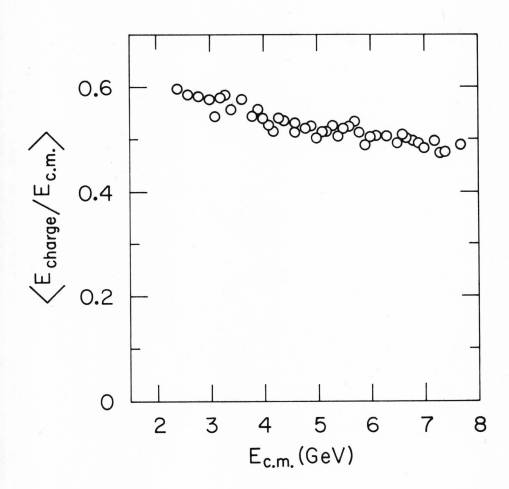

Figure 6 Average fraction of total c.m. energy appearing
in charged particles vs $E_{c.m.}$.

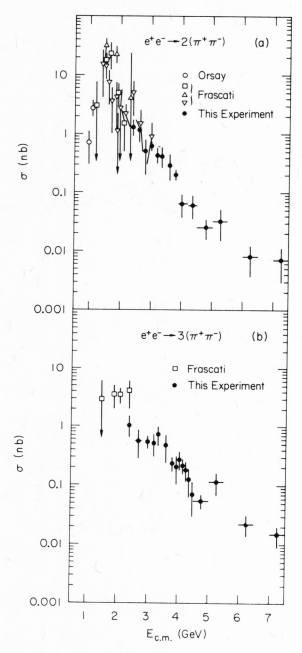

Figure 7 Total cross section for a)$e^+e^- \rightarrow 2(\pi^+\pi^-)$ and
b)$e^+e^- \rightarrow 3(\pi^+\pi^-)$ vs $E_{c.m.}$.

of pions.

Figure 8 shows the mass spectrum of $\pi^+\pi^-$ pairs
in $2(\pi^+\pi^-)$ and $3(\pi^+\pi^-)$ final states for a sample of
the data. The raw events are histogrammed directly
without any corrections. In the $2(\pi^+\pi^-)$ data, there is
a strong ρ signal and a strong f signal. Indeed, the
mass spectrum is fitted by assuming the final states
are entirely $\rho\pi\pi$ or $f\pi\pi$, with a crude ratio
$\rho\pi\pi/f\pi\pi \approx 1.9\pm0.5$, ignoring interference and correlations
between ρ and f. The mass spectrum for the $3(\pi^+\pi^-)$ data
fits well the assumption that all final states are
$\rho\pi\pi\pi\pi$. No f signal is observed.

Figure 9 is a scatterplot of the mass of one $\pi^+\pi^-$
pair versus that of the other $\pi^+\pi^-$ pair in $2(\pi^+\pi^-)$
final states, plotted such the $M_1 > M_2$. We see a definite
clustering at $M_1 \approx M_f$, $M_2 \approx M_\rho$, indicating the presence
of the exclusive channel $e^+e^- \rightarrow \rho f$.

III. ψ' DECAYS

We have a smattering of new results to report on
ψ' decays. We list the known decay modes of the ψ' and
their branching ratios in Table II. We see that almost
40% of the ψ' decays are not directly accounted for.
We can partially account for them indirectly as follows.

1. A few direct decays of the ψ' into "ordinary"
hadrons have been observed. Typically the partial widths
have a ratio

$$\frac{\Gamma(\psi' \rightarrow f)}{\Gamma(\psi \rightarrow f)} \approx 1/3 \left(\frac{BR(\psi' \rightarrow f)}{BR(\psi \rightarrow f)} \approx 1/10 \right),$$

compared to the corresponding partial widths in ψ decay.
Here f represents a specific final state of ordinary

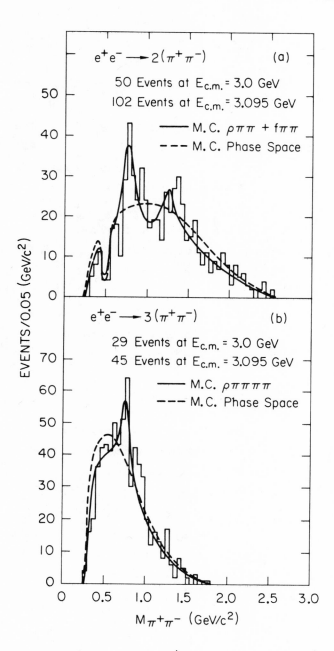

Figure 8   Effective mass of $\pi^+\pi^-$ pairs in a) $e^+e^- \to 2(\pi^+\pi^-)$
and b) $e^+e^- \to 3(\pi^+\pi^-)$.

hadrons. This makes it exceedingly

TABLE II
Decay Modes of the ψ'

Mode	Branching Ratio (%)	Comments
e^+e^-	0.97 ± 0.16	} μ-e Universality assumed
$\mu^+\mu^-$	0.97 ± 0.16	
$\psi\pi^+\pi^-$	32 ± 4	
$\psi\eta$	4.3 ± 0.8	
$\psi\gamma\gamma$	3.6 ± 0.7	via an intermediate state
$\psi\pi^\circ\pi^\circ$	17 ± 3	} includes all $\psi'\to\psi$+neutrals not otherwise identified
ψ anything	57 ± 8	includes all ψ channels above
$p\bar{p}$	0.032 ± 0.014	
$2\pi^+2\pi^-\pi^\circ$	0.35 ± 0.15	
$\pi^+\pi^-K^+K^-$	~0.05	

difficult to observe individual exclusive states in ψ' decay. However, if we assume that the above ratio holds for all decays into ordinary hadrons, we can extimate

$$\frac{\Gamma\psi'\to\text{ordinary hadrons}}{\Gamma\psi'\to\text{all}} \approx 10\%.$$

2. Radiative decays of the ψ' to high mass C even states (χ) have been observed. Events where $\chi\to\psi\gamma$ have already been included in $\psi'\to\psi$ decays. We have also observed events where $\chi\to2(\pi^+\pi^-)$, $3(\pi^+\pi^-)$, $\pi^+\pi^-K^+K^-$, $\pi^+\pi^-$, and K^+K^-. Making reasonable assumptions about the branching ratios of the χ states into these states, we

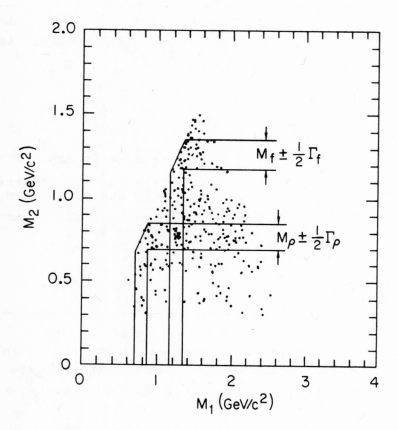

Figure 9 Scatterplot of two $\pi^+\pi^-$ effective masses in
$e^+e^- \to 2(\pi^+\pi^-)$.

can estimate very roughly

$$\sum_{\chi \text{ states}} \frac{\Gamma(\psi' \to \chi)}{\Gamma(\psi' \to \text{all})} \cdot \frac{\Gamma(\chi \to \text{hadrons})}{\Gamma(\chi \to \text{all})} = 5\%.$$

Even so, ~25% of ψ' decays remain unaccounted for.

It has been suggested that the ψ' could decay often into channels such as $\omega\eta$, $\omega\eta'$, or $\omega\chi(2.8)$,[7] channels in which there are two or more neutrals in the final state. These decay modes can be detected due to the electromagnetic decay $\omega\to\pi^+\pi^-$, which occurs with a branching ratio of 1.3%. Since the decays $\psi'\to\rho\eta$, $\rho\eta'$, or $\rho\chi(2.8)$ are forbidden by isospin conservation, the $\omega\to\pi^+\pi^-$ signal should be free of ρ interference. We look for peaks in the $\pi^+\pi^-$ mass spectrum at the ω mass. To increase the sensitivity of the search we first make cuts on the total momentum of the pion pair, since for each channel the ω momentum is unique. We see no evidence for $\omega\eta$, $\omega\eta'$, or $\omega\chi(2.8)$. The 90% c.l. upper limits on branching ratios are

$$\psi'\to\omega\eta < 2.2\%,$$
$$\psi'\to\omega\eta' < 4.3\%,$$
$$\psi'\to\omega\chi(2.8) < 8.6\% .$$

The upper limits are crude, but rule out some models.

The analysis of the decay $\psi'\to\psi\eta$ has now been completed. The following results have been obtained:

$$\frac{\Gamma\psi'\to\psi\eta}{\Gamma\psi'\to\text{all}} = 4.3 \pm 0.8\%,$$

$$\frac{\Gamma\psi'\to\psi\pi^\circ + \Gamma\psi'\to\psi\gamma}{\Gamma\psi'\to\text{all}} < 0.15\% \text{ at } 90\% \text{ c.l.} \quad .$$

If we assume all decays of the type $\psi' \to \psi$ + neutrals other than $\psi' \to \psi\eta$ or $\psi' \to \psi\gamma\gamma$ are $\psi' \to \psi\pi^o\pi^o$, we obtain

$$\frac{\Gamma\psi' \to \psi\pi^o\pi^o}{\Gamma\psi' \to \psi\pi^+\pi^-} = 0.53 \pm 0.06,$$

consistent with the assignment of I=0 to the $\pi\pi$ system. We thus conclude that the ψ' has I=0 and G=-1, and that all $\psi' \to \psi$ decays have now been accounted

TABLE III

Properties of χ States Seen by SPEAR

Mass	Decay Modes Seen	Comments
$\chi(3410)$	$\pi^+\pi^-$, K^+K^-, $2(\pi^+\pi^-)$, $3(\pi^+\pi^-)$, $\pi^+\pi^-K^+K^-$	$J^{PC}=0^{++},2^{++}$, etc.
$\chi(3530)$	$2(\pi^+\pi^-)$, $3(\pi^+\pi^-)$, $\pi^+\pi^-K^+K^-$	Broad state - probably two or more narrow states $\chi(3510)$ and $\chi(3550)$
$P_c(3500$ or $3270)$	$\gamma\psi$	Probably identical with $\chi(3510)$

for to the level of one or two percent. The properties of the χ states observed in $\psi' \to \chi\gamma$ are detailed in Table III.

IV. UPPER LIMITS ON HIGH MASS RESONANCES

The results presented here are not new, but take one new importance with the possible resonance at 6 GeV seen by Lederman.[8] Thus, they bear repeating here. Table IV indicates the experimental upper limits on narrow resonances as a function of c.m. energy. The limits are expressed in terms of the integrated cross section $\int\sigma(E)dE$ expressed in nb·MeV. Narrow here means having a width less

than about 1 MeV, which is the apparent width of a narrow
resonance due to the energy spread in the beam. The limits
on wider resonances have not been worked out in detail,
but at worst our upper limits on integrated cross section
increase linearly with the width of the state.

TABLE IV

Results of the search for narrow resonances. Upper limits
(90% confidence level) for the radiatively corrected integrated cross section of a possible narrow resonance.
The width of this resonance is assumed to be small compared to the mass resolution.

Mass Range (GeV)	Limit on $\int \sigma_H dE_{c.m.}$ (nb MeV)
$3.20 \rightarrow 3.50$	970
$3.50 \rightarrow 3.68$	780
$3.71 \rightarrow 4.00$	850
$4.00 \rightarrow 4.40$	620
$4.40 \rightarrow 4.90$	580
$4.90 \rightarrow 5.40$	780
$5.40 \rightarrow 5.90$	800
$5.90 \rightarrow 7.60$	450

REFERENCES

1.   J.-E. Augustin et al., Phys. Rev. Lett. 34, 233 (1975).

2.   The collaborators in this experiment are G. S.
     Abrams, A. M. Boyarski, M. Breidenbach, F. Bulos,
     W. Chinowsky, G. J. Feldman, C. E. Friedberg,
     D. Fryberger, G. Goldhaber, G. Hanson, D. L. Hartill,
     J. Jaros, B. Jean-Marie, J. A. Kadyk, R. R. Larsen,
     D. Lüke, V. Lüth, H. L. Lynch, R. Madaras, C. C.
     Morehouse, K. Nguyen, J. M. Paterson, M. L. Perl,
     F. M. Pierre, T. P. Pun, P. Rapidis, B. Richter,
     B. Sadoulet, R. F. Schwitters, J. Siegrist, W.
     Tanenbaum, G. H. Trilling, F. Vannucci, J. S.
     Whitaker, F. C. Windelmann, J. E. Wiss.

3.   R. F. Schwitters, Proc. 1975 Int. Symposium on
     Lepton and Photon Interactions at High Energies,
     Stanford University, Stanford, California, 21-27
     August 1975(Stanford Linear Accelerator Center,
     Stanford University, Stanford, California, 1975),
     p.5.

4.   G. S. Abrams, above Proceedings, p.25.

5.   G. J. Feldman, above Proceedings, p.39.

6.   For further information on these events, see M. L.
     Perl, Proc. Canadian Inst. Particle Physics Int.
     Summer School, McGill University, Montreal, 16-21
     June 1975.

7.   The $\chi(2.8)$ has been seen by the DASP collaboration.
     See B. Wiik, Proceedings of 1975 Lepton-Photon
     Symposium (Ref. 3), p. 69.

8.   D. C. Hom et al., submitted to Phys. Rev. Lett.

PRELIMINARY NEUTRAL CURRENT RESULTS
FROM CALTECH-FERMILAB[*]

F. Sciulli

California Institute of Technology

Pasadena, California 91125

The Caltech-Fermilab experiment is completing the
analysis of neutral current data taken about one year ago.
This experiment was carried out about a year ago by the
following people: B. Barish, K. W. Brown, A. Bodek,
D. Buchholz, E. Fisk, G. Krafczyk, F. Merritt, F. Sciulli,
L. Stutte, and H. Suter. The final results should be
forthcoming in the next few weeks. This communication is
a status report on that analysis.

The data were taken in the Fermilab narrow band
beam set to pion and kaon secondaries of mean energy ±170
GeV. The distributions in total energy for the charged
current events, shown in figure 1, reflect the dichromatic
beam structure with peak energies of about 50 and 150
GeV. The comparison of ν and $\bar{\nu}$ neutral current rates is
crucial for the determination of the Lorentz-coupling, so
it is imperative that the data of each type contain little

---

[*] Work supported by the U.S. Energy Research and Development Administration. Prepared under Contract E(11-1)-68
for the San Francisco Operations Office.

CITF DATA — DISTRIBUTIONS IN TOTAL OBSERVED
ENERGY NARROW BAND BEAM CHARGED — CURRENTS

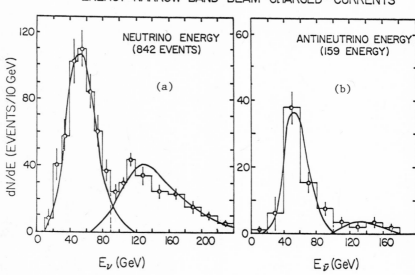

Figure 1

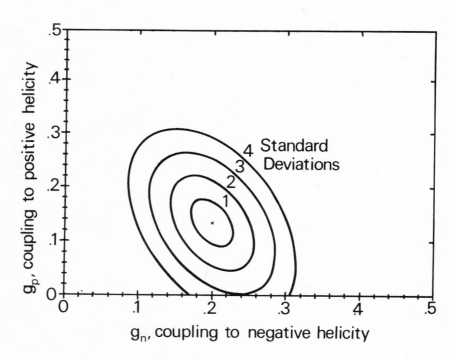

Figure 2

or no contamination from the neutrino of opposite helicity.
Neutrinos from pion and kaon decay in the decay pipe have
no such contamination, because the narrow band beam will
not pass any mesons of the wrong charge. Neutrinos from
decays upstream of the decay pipe (wide-band background)
are not so clean, however. In this experiment, we measured
separately this wide-band background after closing a slit
at the decay pipe entrance. This measured background was
empirically subtracted from the data.

The data were separated into neutral (NC) and
charged-current (CC) components in the usual way, i.e.
by calling events with penetrating muons (>1.6m steel)
charged current events. This procedure leaves some component of CC as background in the NC sample. This question
will be addressed later.

We proceed to fit the distributions in hadron energy,
$E_h = y E_\nu$, to the general form:

$$\frac{dN^\nu}{dy} = \frac{G^2 ME_\nu}{\pi} F_\nu [g_n + g_p (1-y)^2] \ , \qquad (1a)$$

$$\frac{dN^{\bar\nu}}{dy} = \frac{G^2 ME_{\bar\nu}}{\pi} F_{\bar\nu} [g_n (1-y)^2 + g_p] \ , \qquad (1b)$$

where parameters g_n, g_p are the couplings of negative and
positive helicity measured relative to the charged current
Fermi constant, G. These values contain contributions to
g_n from V-A neutrino-quark scattering as well as V+A
neutrino-antiquark scattering. For g_p, V+A neutrino-quark
scattering and V-A neutrino antiquark scattering contribute.

Aside from the usual statistical errors in the lefthand side of equation 1, coming from the limited data
sample and the empirical subtraction already mentioned,
there are two pieces of information that are required for

such a fit:

(1)   The values of the flux parameters F_ν, $F_{\bar\nu}$

(2)   Correction for the CC events remaining in the
      neutral current sample. These are primarily
      events of low muon energy, and therefore large

We have used the charged-current data to obtain the necessary information for both cases.

The first question can be resolved by using the CC
data at small y. If charge symmetry is correct at low
invariant masses (as seen at Gargamelle), then the value
of $dN/dy\,\big|_{y=0}$ for charged-currents measures precisely the
factor $\frac{G^2ME}{\pi}\nu\,F_\nu$. In practice, of course, we must extra-
polate with some model from finite y (y<0.2) to zero y.
However, the dependence of this factor on the detailed
model assumed for the charged currents is very small in
comparison to its statistical error.

The second question is somewhat more complicated and
on its surface, more dependent on the detailed model assu:
ed for charged currents. This is true because the correc
tion comes at large values of y, where production of new
hadrons is expected to be most dominant. Experimentally,
there have been reports of energy dependence in the value
of <y> for anti-neutrinos. In our own charged current
data, there are hints of some such energy dependence at
large y. Unfortunately, the charged-current anti-neutrin
data is too statistically limited to unambiguously claim
such a break-down in scaling. The data can be fit with
an integrated $\bar{Q}$ fraction α in the range: $0.1<\alpha<0.3$. On
the other hand, the data can be fit somewhat better by
models incorporating an increase in events at large y for
antineutrinos as the energy is increased.

The only question of major importance remaining to
be resolved for the neutral currents is whether the sub-

traction of large y CC events from the NC data sample is
dependent on the CC model to any substantial extent.   As an
example, figure 2 shows the probability contours for g_n
vs. g_p for fits to the neutral current data using the
following model for the CC subtraction: scaling with
integrated $\bar{Q}$ fraction, α_{cc}=0.17.   The best values are

$$g_n^{nc} = .200 \pm .029 \, ,$$

$$g_p^{nc} = .134 \pm .044 \, ,$$

where the quoted errors include the statistical errors
on the data points and on the flux factors, but do not
include systematic errors on the subtraction of CC background.

The more directly interpretable parameters are the
neutral current couplings, g_-^{nc} and g_+^{nc}, for V-A and V+A
currents, respectively.   These are related to the measured
numbers as follows:

$$g_- = \frac{(1-\alpha)g_n - \alpha g_p}{1-2\alpha} \quad ,$$

$$g_+ = \frac{(1-\alpha)g_p - \alpha g_n}{1-2\alpha} \quad ,$$

where α^{nc} has not been measured.   For example, a purely
vector interaction with equal amounts of V-A and V+A has
$g_-=g_+$, whereas the Weinberg model gives the predictions
in terms of the Weinberg angle, x= $\sin^2\theta_w$: $g_-=\frac{1}{2}-x+\frac{5}{9}x^2$;
$g_+=\frac{5}{9}x^2$.

Figures 3, 4, 5 show the values for g_+, g_- obtained
with the assumptions α = .06, .17, .34, respectively, as
examples.   The data shown are definitely inconsistent with

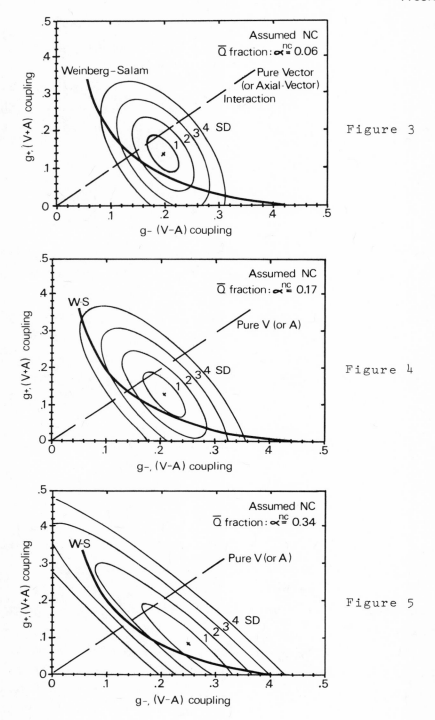

Figure 3

Figure 4

Figure 5

V+A, and about 1 SD statistically either pure V or the Weinberg-Salam assumptions. The data are only consistent with V-A for rather large values of neutral current anti-quark, α.

We are proceeding to examine the effect of various charged current model assumptions on the subtracted background, and therefore on the systematic errors for g_p and g_n. The specific determination of α^{nc} must await experiments capable of determining the x-distribution for neutral currents.

SEARCH FOR NEW PARTICLES AT BNL

Min Chen

Department of Physics & Laboratory for Nuclear Science

Massachusetts Institute of Technology

Should the newly discovered particles J (3.1), ψ (3.7) etc. be explained as particles with some hidden quantum number, e.g. charm, then particles with that explicit quantum number should also exist. The MIT-BNL group has completed a systematic search for such particles in the following reaction:

$$P + Be \rightarrow h^+ + h^- + x$$

$$\rightarrow h^\pm + e^\mp + x$$

$$\rightarrow \mu^\pm + h^\pm + h^\pm + x \quad ,$$

where $h^\pm$ stands for hadrons ($\pi^\pm, k^\pm, p^\pm$) .

(I) In the $h^+ h^-$ mode

A systematic search in the following reactions in the pair mass region 1.2 - 5.5 BeV with a mass resolution of 5 MeV was made:

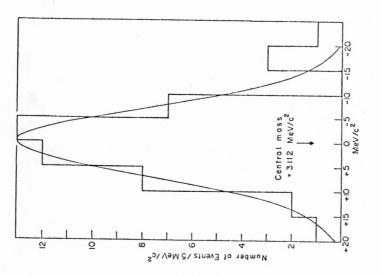

Central mass
= 3112 MeV/c²

The width of J

b

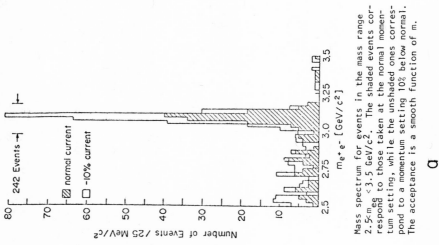

Mass spectrum for events in the mass range
$2.5 < m_{ee} < 3.5$ GeV/c². The shaded events correspond to those taken at the normal momentum setting, while the unshaded ones correspond to a momentum setting 10% below normal.
The acceptance is a smooth function of m.

a

Figure 1   Target vertex reconstruction for hadron pair data.

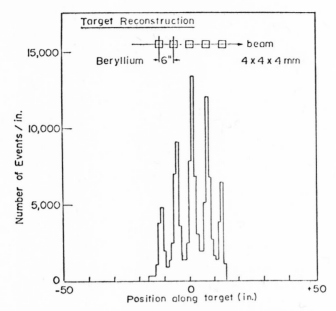

a. Reconstruction of the pair vertex at the target; using information from the proportional chambers. The five pieces of beryllium are seen clearly.

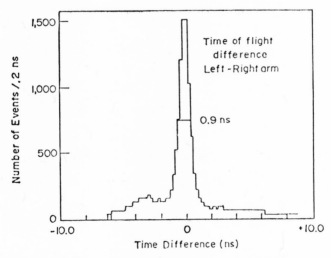

b. Time difference between additional scintillation counters in the left and right arms. The resolution obtained is 0.9ns and little background is present.

Figure 2   Time difference of the two hadrons as measured in the two arms of the spectrometer.

$$p + p \rightarrow \pi^- p + x \qquad\qquad A$$

$$\left.\begin{array}{l} \pi^+ \pi^- + x \\[4pt] \bar{p} p + x \\[8pt] K^- p + x \\[8pt] K^+ \pi^- + x \end{array}\right\} \qquad B$$

$$\left.\begin{array}{l} K^+ K^- + x \\[4pt] K^- \pi^+ + x \\[4pt] K^+ p^- + x \\[4pt] \pi^+ p^- + x \end{array}\right\} \qquad C$$

The experiment set up for this search is very similar to the original $e^+ e^-$ experiment where the J particle was first observed in August 1974. However, since there were many more hadrons than electrons the random accidentals were more serious. To reduce the accidentals to the minimum a new target system was put in. It consisted of 5 pieces of 4 mm x 4 mm x 4 mm Be target each separated by six inches. The targets were supported by thin piano wires. This arrangement enabled one to locate the point of intersection between two trajectories. Comparison with the target location enabled us to reduce random accidentals (Fig. 1). To further reduce accidentals additional scintillation counters were installed to tighten the two arm coincidence to 0.9 ns (Fig. 2). Two high pressure (300 psi) Cerenkov counters were installed to identify K's, replacing the shower counters. The counters C_e, C_o were filled with 1 atm isobutane to identify π's. The Cerenkov counters set the mass acceptance to ≈ 1 BeV. In this way all 9 combinations were measured simultaneously.

To avoid systematic errors, 7 overlapping magnet settings
were made for the measurement.

Figures 3 to 5 show the results of some of these
typical reactions. Without acceptance corrections the
yield increases with mass due to an increase in
acceptance. It then decreases due to the decline in
production cross sections. There are no sharp narrow
resonances in any of the 9 reactions. There may of
course be very wide "ordinary" resonances with widths of
300 MeV or more. A search for these depends on exact
calculations of acceptance and as yet have not been
made.

To gain a feeling for the sensitivity of the
measurements, we take the production mechanisms of the
9 reactions to be the same as that assumed for J. From
this we obtain the following table:

$$\text{Sensitivity } (cm^2)$$

$$\text{For Narrow Resonances}$$

$h^+ x^-$ \ m	2.25 GeV	3.1 GeV	3.7 GeV
$\pi^+ K^-$	1×10^{-33}	4×10^{-35}	1×10^{-35}
$K^+ \pi^-$	4×10^{-33}	8×10^{-35}	4×10^{-35}
$p\ p^-$	-	4×10^{-34}	2×10^{-35}
$K^+ K^-$	1×10^{-33}	5×10^{-35}	1×10^{-35}
$\pi^+ \pi^-$	8×10^{-33}	5×10^{-34}	3×10^{-35}

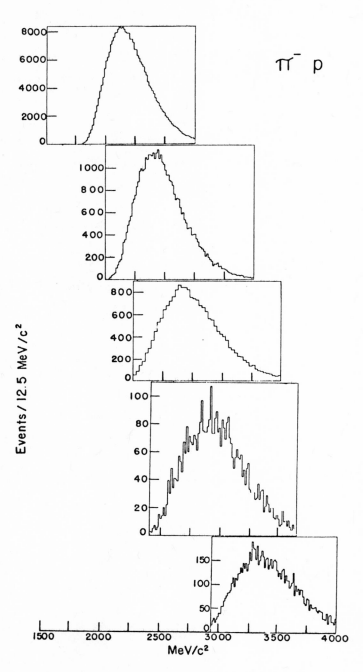

Figure 3   Event distribution of π⁻p.

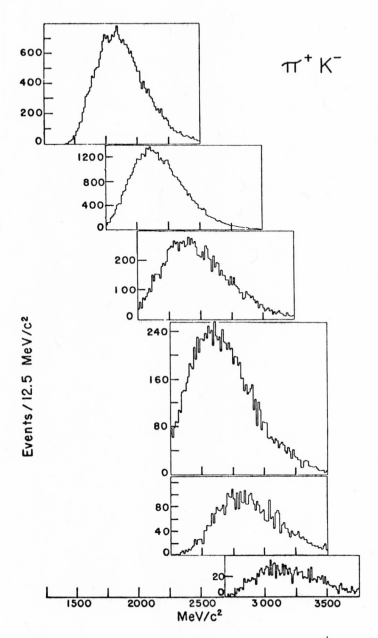

Figure 4   Event distribution of $k^+\pi^-$.

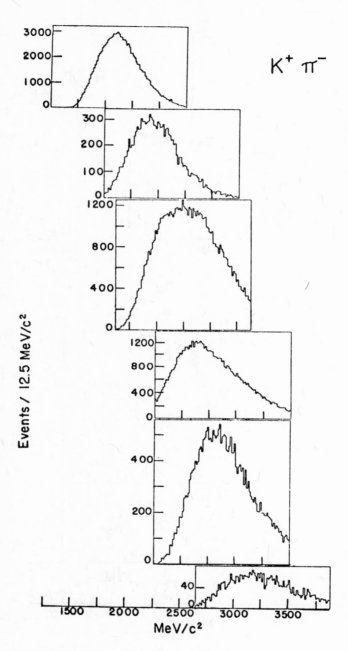

Figure 5   Event distribution of $\pi^{+}K^{-}$.

Whereas the spectra do not show any sharp re-
sonance states, the cross sections $\frac{d\sigma}{dm}$ vs m for groups
A (π^-p), B $(\pi^+\pi^-$, $\bar{p}p$, $K^+\pi^-$, $K^-p)$ and C $(K^+K^-$, $K^+\bar{p}$, $\pi^+\bar{p}$,
$K^-\pi^+)$ do exhibit some simple degeneracies above the mass
of J. The cross sections for each group decrease with
a mass $\approx e^{-5m}$ and differ from each other by an order of
magnitude (Fig. 6).

The cross sections in the pair range from 3 to 5.5
GeV is shown in Fig. 7. This is the energy region where
the ratio $e^+e^- \rightarrow$ hadrons$/\mu^+\mu^-$ increases from 2.5 to 5.
However the pair cross section measured here showed
monotonically decreasing with the same slope e^{-5m} as
in the low mass region.

(II) In the electron-hadron mode

With the right arm of the spectrometer tuned to
detect hadrons, the left arm was converted to detect
electrons (positrons): Both of the Cerenkov counters
were filled with hydrogen gas and the lead glass and
shower counters are put behind C_k to measure the pulse
height of electromagnetic showers.

We scanned through the pair mass region from 1.5 to
3 GeV in three settings. The polarity of the spectrometer was reversed several times to measure both h^+e^-
and h^-e^+ pairs.

Fig. 8 shows the relative timing of the electron
and the hadron as measured in the spectrometer. We see
a clear coincidence signal due to π^+e^- and π^-e^+. Some
typical mass spectra of π^+e^- and π^-e^+ are shown in Fig. 9.
A comparison of the genuine coincidence pairs with random
accidental π e pair shows no difference in the mass
spectra. In other words there is no indication of a
discontinuity in the mass spectra due to some resonance

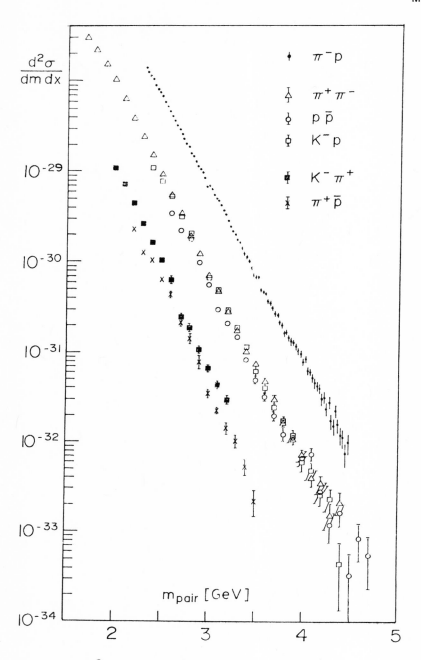

Figure 6 $\dfrac{d^3\sigma}{dp\ dx\ dm}$ for h^+h^- data in the mass region 1.5 to 4 GeV.

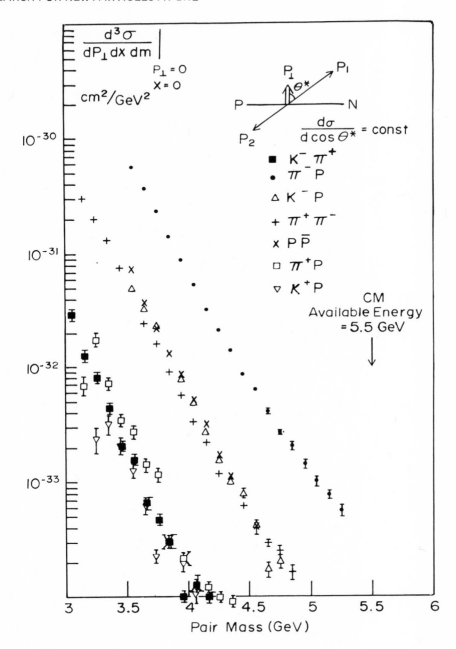

Figure 7 $\dfrac{d^3\sigma}{dp\ dx\ dm}$ for h^+h^- data in the mass region 3 to 5.5 GeV.

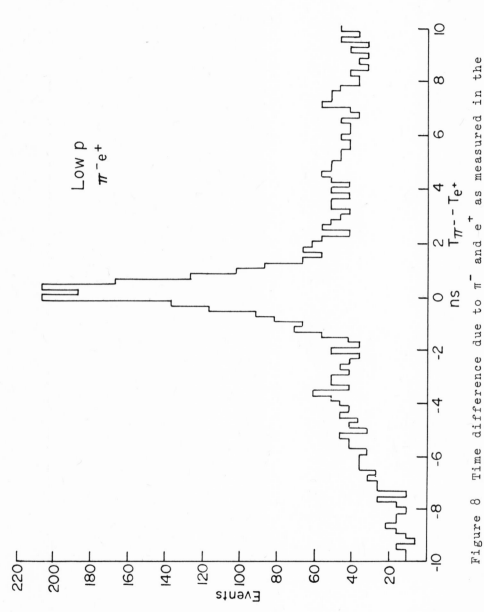

Figure 8   Time difference due to π^- and e^+ as measured in the spectrometer.

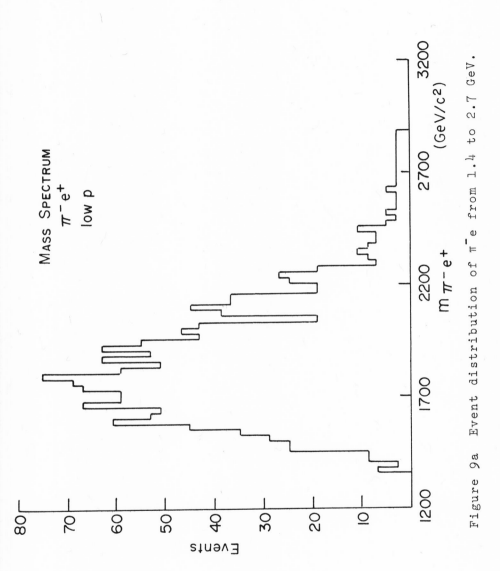

Figure 9a   Event distribution of $\pi^- e$ from 1.4 to 2.7 GeV.

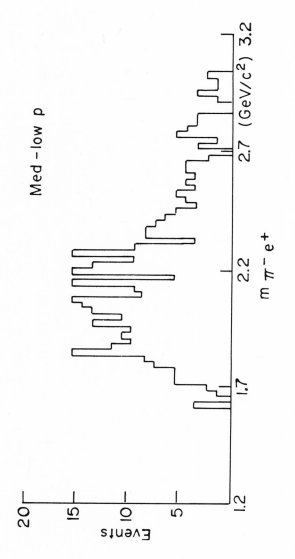

Figure 9b   Event distribution of π^-e from 1.7 to 3 GeV.

decay such as $D \rightarrow e + h + x$.

Another interesting feature is the missing of the genuine coincidence of the signal due to $K^- e^+$ and $\bar{p} e^+$ which were recorded at the same time as the $\pi^- e^+$ events. In the hadron momentum range between 3.0 and 6.0 GeV the upper limit of $K^- e^+$ and $\bar{p} e^+$ are

$$\frac{\sigma(K^- e^+)}{\sigma(\pi^- e^+)} < 1\% \quad ,$$

and

$$\frac{\sigma(\bar{p} e^+)}{\sigma(\pi^- e^+)} < 1\% \qquad (95\% \text{ confidence level}).$$

During the same running periods, 8 $e^+ e^-$ events with mass = 3.1 ± 0.007 were also observed. These events are clearly identified to be the J particle. Therefore the upper limit of the $K^- e^+$, $\bar{p} e^+$ channel can also be normalized to the production cross section of the J. We obtain (90% confidence level)

$$\frac{\int_2^3 \frac{d\sigma}{dm} (p + Be \rightarrow K^- + e^+ + x) \times (\text{Acceptance}) \, dm}{\sigma_J (p + Be \rightarrow J + x) \times \text{Acceptance} \quad \underset{e\ e}{\vdash}} < 7$$

and

$$\frac{\int_{2.3}^{3.3} \frac{d\sigma}{dm} (p + Be \rightarrow \bar{p} + e^+ + x) \times \text{Acceptance} \, dm}{\sigma_J (p + Be \rightarrow J + x) \times \text{Acceptance} \quad \underset{e\ e}{\vdash}} < 1 \quad ,$$

where m is in GeV.

(III) In the $\mu h^+ h^-$ mode

As we discussed in (I) that charm was not observed in the $h^+ h^-$ mode. There are however two possible ex-

planations that charmed particles may be produced
abundantly yet there is no significant signal in the
h^+h^- mode: 1) The branching ratio of two body decay
mode may be very low; 2) The noncharmed h^+h^- cross
sections are much larger than the charmed ones such
that the latter is swamped.

In order to enchance the signal due to charm,
the following observations were made:

1)   In hadron interactions charmed particles are
produced in pairs in order to conserve quantum number.

2)   There are theoretical speculations that the
leptonic or semileptonic decay modes are of substantial
size (8%).

Therefore an event selection for h^+h^- with a μ as a
fragment from the other charm partner detected will be a
much more selective signature of charm.   One such possibl
reaction is

$$p + p \rightarrow D + \bar{D} + x$$

$$\quad\quad\quad\quad\mathrel{\llcorner} K\pi \text{ or } Kp \text{ if Baryon}$$

$$\quad\quad\mathrel{\llcorner} K\ \mu\nu \text{ or } \Lambda\mu\nu$$

Muon Detector

One of the unique features of the MIT-BNL double
arm spectrometer is that it can withstand a very intense
beam.   Hence, the muon detector should have the following
properties:

1)   Operate under as intense hadron fluxes as the
double arm spectrometer can.

2)   Detect muons with CM momentum $p^* \geq \frac{M_D}{3} = 700$ **MeV**,
where the mass of charmed mesons is expected to be $M_D \sim 2$ G

3)   Cover a large solid angle.

4)   Absorb kaons and pions right next to the target

to reduce muon background from decays.

Figure 10 shows the plan and side view of the
detector. Tungsten blocks are piled around the beam
1" away from the target. A sealed one cubic meter
box of uranium surrounds the tungsten collimator to
further absorb the hadron shower. Beyond the uranium
is one meter of lead concrete sandwich reducing the
number of neutrons to a tolerable level. The detector
covers a solid angle of 2 sr. and accepts muon momenta
greater than 2.8 GeV in the laboratory or .7 GeV in the
center of mass.

The coincidence between two banks of scintillation
counters registers muon candidates. With 4×10^{10}
protons per pulse incident upon a 2% collision length
target, we established that the single rate of all
counters is less than 4 MC. The coincidence rate of
the two scintillation counters is about 100 KC.

We have observed muons in coincidence with particles detected by the MIT-BNL double arm spectrometer.
Figure 11 shows the pulse height distribution of the
muon scintillation counters. Figure 13 shows the
coincidence of two muon counters and Fig. 14 displays
the triple coincidence between a muon and particles in
each of the spectrometer arms at high beam intensities.

We estimate a hadron punch through of less than
10^{-4} and approximately 5 in 10^{4} pions decaying to
contaminate the muon rate.

In the center of mass frame the detector covers a
solid angle of roughly 2 s.r. The c.m.s. production
angle of detected muons, θ^{*}, ranges from 60° - 120°, with
a minimum required momentum, p*, of approximately 600
MeV. The acceptance of muons as functions of different
values of $p_{\perp}^{*}$ and $p_{\parallel}^{*}$ in the cm system is shown in Fig. 15.

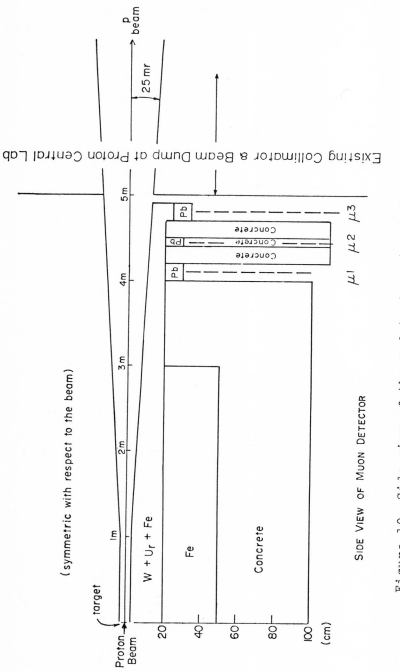

Figure 10   Side view of the μ detector siguated at the front
of the spectrometer.

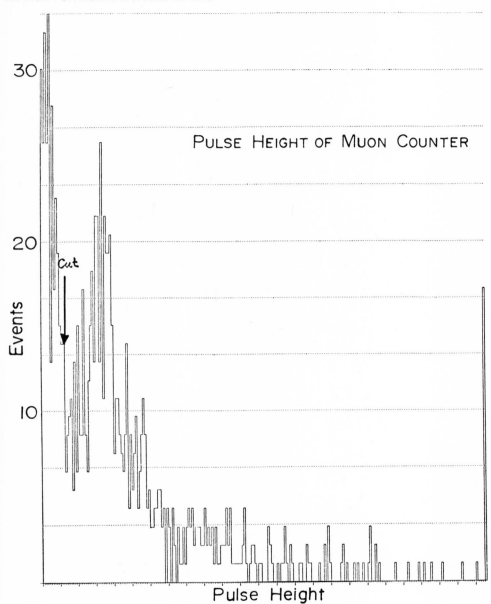

Figure 11 Pulse height distribution of particles going
           through the muon scintillation counter. One
           sees a peak due to relativistic charged part-
           icles and a tail due to low energy neutron or
           photon conversion.

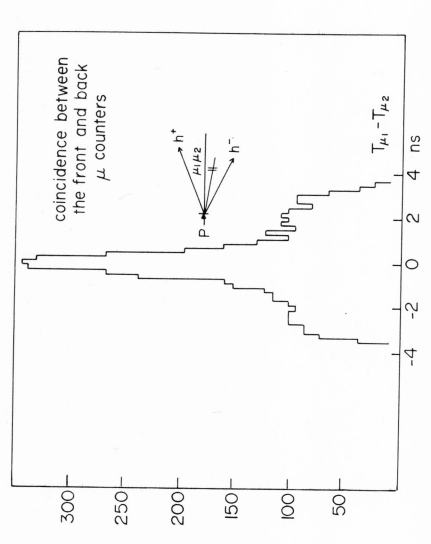

Figure 12   Time difference between the two muon scintillation counters.
The peak corresponds to particles going through both
counters at once.

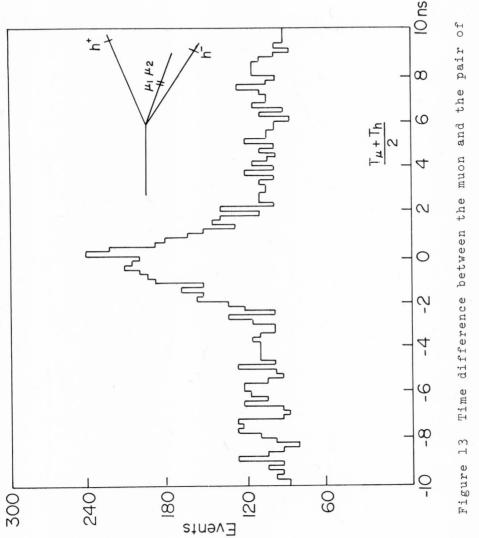

Figure 13   Time difference between the muon and the pair of
            hadrons.

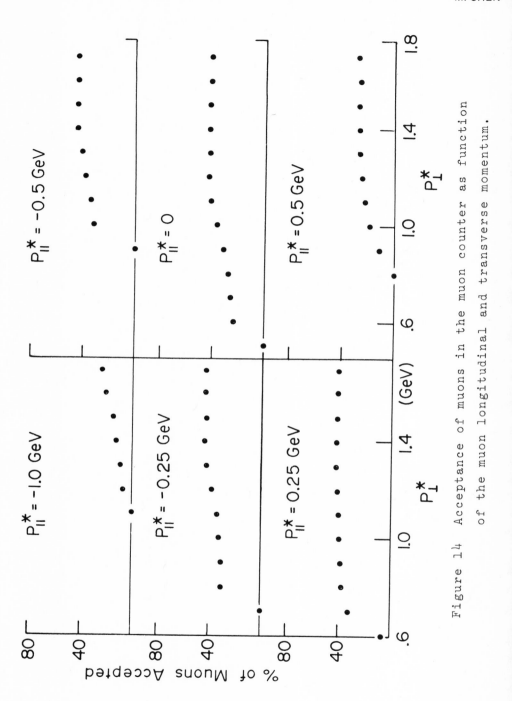

Figure 14 Acceptance of muons in the muon counter as function of the muon longitudinal and transverse momentum.

The event rate with the muon trigger requirement is a
factor of a thousand lower than that of the double arm
spectrometer alone.

The $\pi^+\pi^-$, π^-p mass spectra with μ are shown in
Fig. 15. The corresponding mass spectra without μ in
coincidence are shown in Fig. 16. The ratio of the
$\pi^+\pi^-$, π^-p mass spectra with/without μ is flat and
structure less. The same ratio for $K^-\pi^+$ channel is
shown in Fig. 17. Again it is flat. In conclusion we
see no structure in the reactions

$$
\begin{aligned}
p + Be &\rightarrow \pi^- + p + \mu + x \\
&\rightarrow \pi^+ + \pi^- + \mu + x \\
&\rightarrow p + \bar{p} + \mu + x \\
&\rightarrow K^- + p + \mu + x \\
&\rightarrow \pi^+ K^- + \mu + x \\
&\rightarrow K^+ K^- + \mu + x \\
&\rightarrow K^+ \bar{p} + \mu + x \\
&\rightarrow \pi^+ \bar{p} + \mu + x \quad ,
\end{aligned}
$$

in the pair mass range from 1.5 to 2.6 GeV. There is
some structure observed in the $K^+\pi^-$ channel but it is
still statistically not convincing. We are in the
process of investigating this channel in greater detail
and statistics and hope to have results soon.

If the observed direct μ's are all originated from
charmed particle decay, one would expect a cross section
due to charm about 10^{-29} cm^2. The upper limit we can
set for the channel $K^-\pi^+\mu$ is about 10^{-31} cm^2/GeV in the
mass range $m_{K\pi} = 1.5$ to 2.5 GeV, which shows that

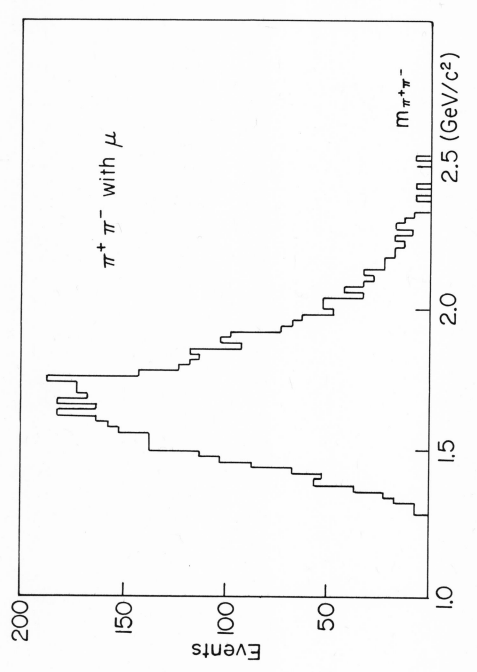

Figure 15a Event distribution of $\pi^+\pi^-$ with a muon detected in

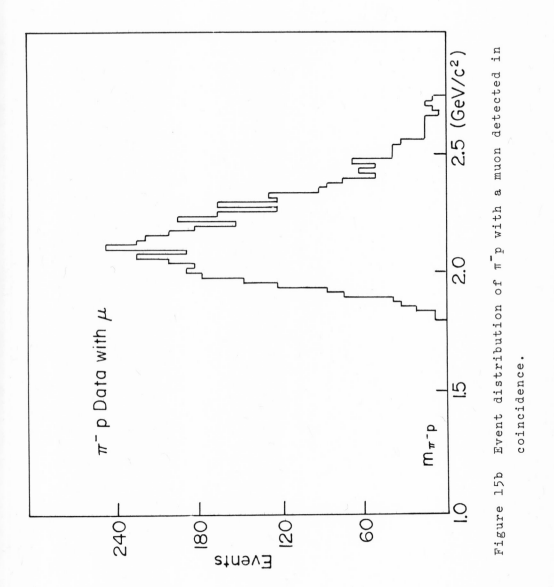

Figure 15b   Event distribution of π^-p with a muon detected in
coincidence.

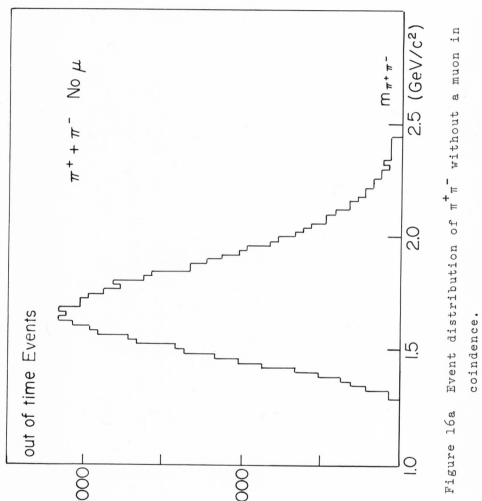

Figure 16a Event distribution of $\pi^+\pi^-$ without a muon in coindence.

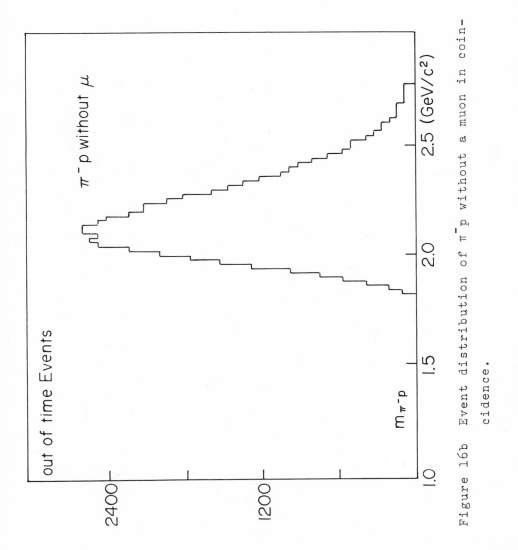

Figure 16b   Event distribution of π^-p without a muon in coincidence.

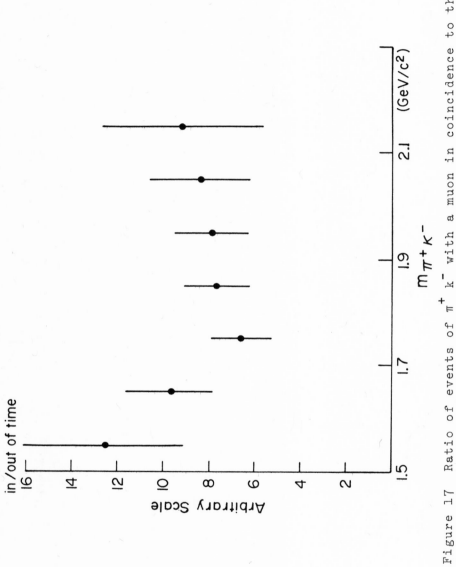

Figure 17 Ratio of events of $\pi^+ k^-$ with a muon in coincidence to the events without muon in coincidence as function of the pair hadron mass.

that $(K^-\pi^+\mu) + x$ is not a significant decay mode of
the pair of charmed particles.

Dr. Zee is conversing with Professor Murray Gell-Mann and
and the editor, Professor Arnold Perlmutter

NEW VISTAS IN NEUTRINO PHYSICS*

A. K. Mann

Department of Physics

Univ. of Pennsylvania, Philadelphia, Penn. 19174

Recent data on the high-y anomaly in inelastic $\bar{\nu}_\mu$ -nucleon scattering and on the ratio of cross sections $\sigma_c^{\bar{\nu}}/\sigma_c^{\nu}$ are presented and interpreted in terms of new hadron production.

There exist data on high energy inelastic ν- and $\bar{\nu}$-nucleon collisions that lead to a single final state muon which are not easily understood by extrapolation to higher energy of our present knowledge of semi-leptonic weak processes at lower energy.[1,2] The reactions of interest are

$$\nu_\mu + \text{nucleon} \rightarrow \mu^- + H$$

and

$$\bar{\nu}_\mu + \text{nucleon} \rightarrow \mu^+ + H,$$

where H is any assembly of hadrons. What was observed - and since named the high-y anomaly - was that the in-

*Work supported in part by the U. S. Energy Research and Development Administration.

elasticity distribution for $\bar{\nu}$ collisions, i.e., the
distribution in the variable $y = E_H/E_{\bar{\nu}}$, is much
flatter than expected[1] at high $E_{\bar{\nu}}$. Furthermore, the
invariant mass (W) distributions of the recoiling hadron
system in the $\bar{\nu}$ scattering process also departed from
the expected form.[2] Both effects exhibited a dynamic
energy threshold in the vicinity of $E_{\bar{\nu}} \approx 30$ GeV. These
results - manifesting effective violations of scale
invariance and charge symmetry invariance - strongly
suggested new particle production by $\bar{\nu}$; no evidence
for or against new particle production by ν was obtained
from them.

Also reported was additional stronger evidence for
new particle production in both ν and $\bar{\nu}$ interactions
which consisted of events in which two muons (dimuons)
were observed in the final state.[3,4,5] These reactions
are empirically described by

$$\nu_\mu + \text{nucleon} \rightarrow \mu^- + \mu^+ + \nu_\mu + H \, ,$$

$$\bar{\nu}_\mu + \text{nucleon} \rightarrow \mu^- + \mu^+ + \bar{\nu}_\mu + H,$$

and show an energy threshold similar to that observed in
the $\bar{\nu}$ single muon data. In the main, the source of the
dimuons appears not to be the production and decay of
either neutral or charged intermediate vector bosons
or neutral heavy leptons. Nor are the dimuons due to
a direct four-fermion process.

A likely common explanation of both single muon and
dimuon data is the production of one or more massive,
short-lived new hadrons that decay weakly and therefore
carry a new quantum number.[2,4,5,6,7] It is of interest,
however, to inquire whether possible systematic errors
in resolution or in detection efficiency might be responsible for the high-y anomaly and the unexpected W-

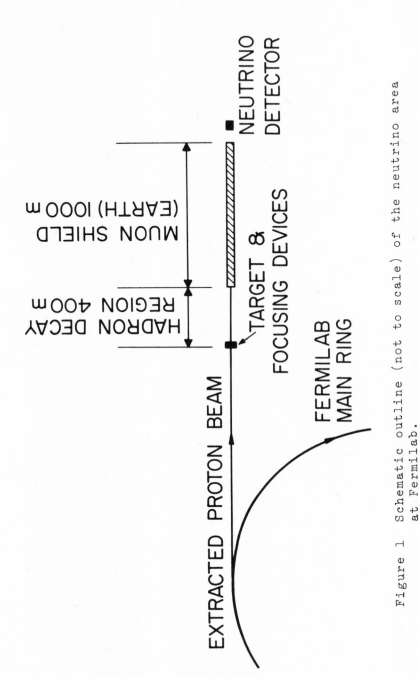

Figure 1 Schematic outline (not to scale) of the neutrino area at Fermilab.

distributions in the $\bar{\nu}$ single muon data. This would, of course, call into question the common explanation of the single muon and dimuon data. In this paper, therefore, we address the subjects of resolution and detection efficiency in the HPWF neutrino apparatus, and utilize larger samples of ν and $\bar{\nu}$ data to repeat the previous search for energy dependent anomalous distributions.[8] In short, we ask the question: does the high-y anomaly stand up under closer scrutiny and more data? With these larger samples we also consider the ratio of total cross sections $\sigma_c^{\bar{\nu}}/\sigma_c^{\nu}$ for the reactions with single final state muons.[9]

The geometry of the neutrino area at Fermilab is shown in Fig. 1. The beams in that area vary in type depending on the nature of the focusing of the secondary hadrons that decay to produce the ν and $\bar{\nu}$ beams. In Fig. 2a are shown ν energy spectra produced by 400 GeV protons for different types of secondary hadron focusing.[10] Similar spectra for $\bar{\nu}$ are shown in Fig. 2b. Notice that the bare target spectra, i.e., spectra resulting from the absence of focusing of the secondary hadrons, are lowest in intensity and average energy. The ν and $\bar{\nu}$ spectra produced by magnetic horn focusing and by quadrupole triplet focusing differ significantly in their hardness; quadrupole triplet focusing leads to appreciably harder ν and $\bar{\nu}$ spectra. In the data presented below, there are samples taken with both quadrupole triplet and magnetic horn focusing. These samples yield essentially the same distributions in the dynamic variables under consideration and serve, therefore, to demonstrate that neither the experimental resolution nor the detection efficiency depended in any unexpected way on the nature of the incident ν and $\bar{\nu}$

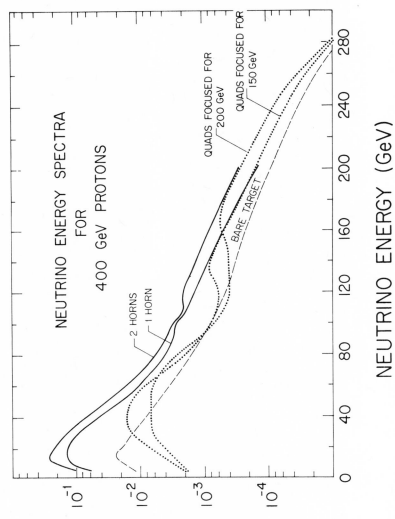

Figure 2a   Neutrino spectra produced by 400 GeV protons with
different focusing of the secondary mesons.

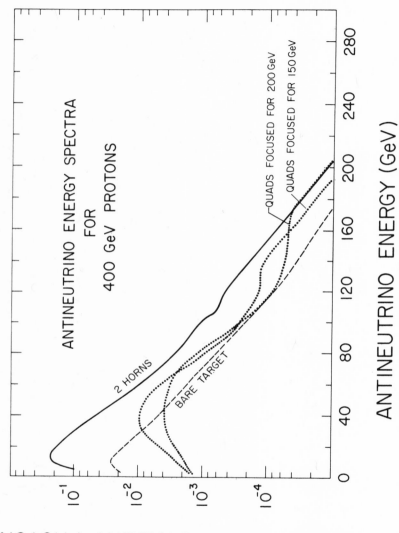

Figure 2b   Antineutrino spectra produced by 400 GeV protons with
different focusing of the secondary mesons.

beams.

In Fig. 3 is shown in schematic outline the HPWF
experimental apparatus which has been described in
detail elsewhere.[11,12] The semiempirically determined
resolution function for that apparatus is shown in
Fig. 4. Here we give the resolution for the independent dimensionless variables $x = Q^2/2mE_H$ and y, where
$Q^2 = 4E_\nu E_\mu \sin^2(\theta_\mu/2)$, E_H is the energy of the hadron
cascade, $E_\nu = E_H + E_\mu$, and m is the nucleon mass. Using
the information in Fig. 4 we have made Monte Carlo
calculations of the patterns of migration of events in
the x-y plane due to resolution smearing. For example,
we calculate in Fig. 5a the migration pattern of events
that start in the limited region 0.1 < x < 0.2,
0.1 < y < 0.2. Observe that events move primarily in
x, and to a much lesser extent in y. This is also
borne out in Fig. 5b where the unmigrated events start
from the bin 0.1 < x < 0.2, 0 < y < 0.1. Assuming a
trial function in x and y of the form $F_2(x) (1-y)^2$,
where $F_2(x)$ is obtained from inelastic electron
scattering data, we have calculated the effect of
resolution smearing on the distributions in y for $\bar\nu$
in two regions of $E_{\bar\nu}$. This is shown in Fig. 6, where
the y-distributions are integrated over the region
$0 < x \leq 0.6$, because the resolution in x becomes poor
at x > 0.6 (see Fig. 4), and also because there are
relatively few events in the region x > 0.6 in the
experimental distributions presented below. Fig. 6
demonstrates that, apart from a small (<10%) diminution
of events at low-y, the trial function $(1-y)^2$ is
essentially unmodified by resolution smearing. Migration
patterns constructed for different trial distribution,
different energy intervals, different regions of x,

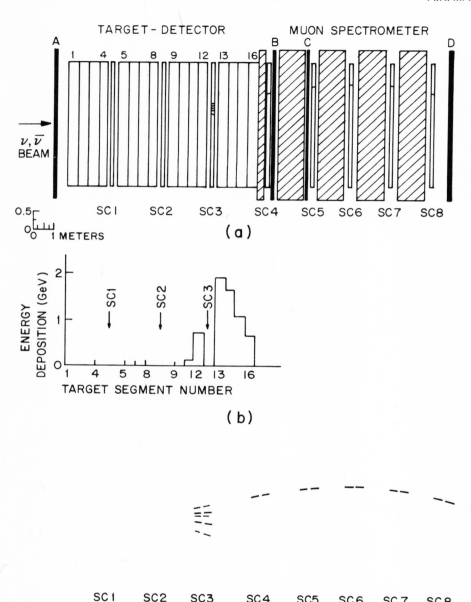

Figure 3 Schematic outline of the Harvard-Pennsylvania-
Wisconsin-Fermilab neutrino detector.

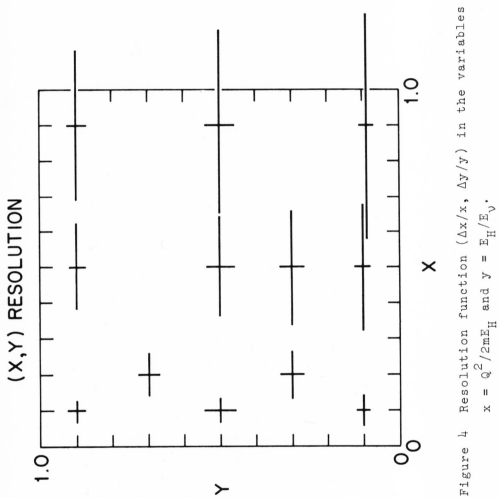

Figure 4   Resolution function $(\Delta x/x,\ \Delta y/y)$ in the variables $x = Q^2/2mE_H$ and $y = E_H/E_\nu$.

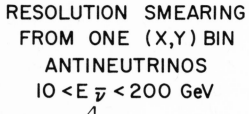

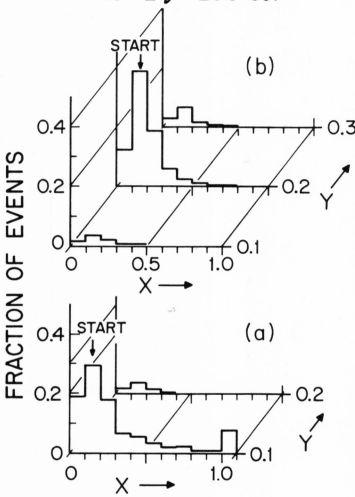

Figure 5 Migration patterns of events starting in a
          single (x,y) bin. (a) Start in
          0.1 < x < 0.2, 0.1 < y < 0.2; (b) start in
          0.1 < x < 0.2, 0 < y < 0.1.

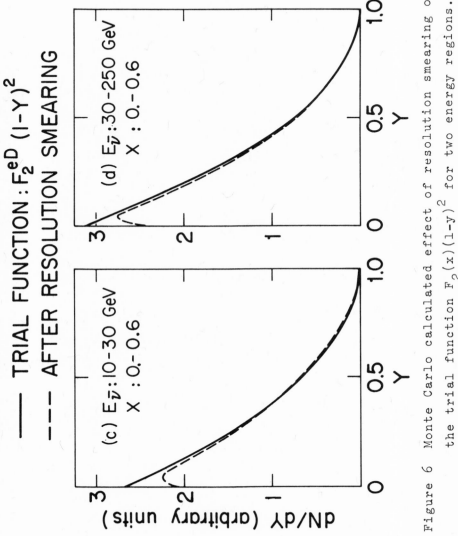

Figure 6   Monte Carlo calculated effect of resolution smearing on the trial function $F_2(x)(1-y)^2$ for two energy regions.

and with arbitrary variations of the resolution function of Fig. 4 all show similar behavior, and lead
to the conclusion that resolution smearing is not the
the source of the observed high-y anomaly.

In Fig. 7 are shown the calculated limits on the
geometric detection efficiency of the apparatus (Fig.
3) in the Bjorken x-y space for different $\nu, \bar{\nu}$ energies.
These limits are somewhat arbitrarily set by the requirement that, in any x-bin, the integrated efficiency
in y below the limit is 0.8. Also shown in Fig. 7
(as a sample) is the calculated distribution in geometric detection efficiency ε_μ in the particular energy
interval 30 < $E_{\nu, \bar{\nu}}$ < 70 GeV.   Using detection efficiency calculations represented by Fig. 7, a weighting
factor (ε_μ^{-1}) has been applied to each observed event
to correct the data for detection efficiency in a
model independent fashion. No correction has been
applied for unobserved events outside the angular
acceptance or muon range cut-off of the apparatus because it would involve model dependent assumptions to
correct the data in those regions of very low detection
efficiency. Nor has any correction been made for resolution smearing. In the distributions in y of the
data we have integrated over the interval x $\leq$ 0.6 to
avoid the region of relatively poor detection efficiency
at x > 0.6, which is also the region of poor resolution
as we have seen. No conclusions reached below depend
on that cut in x.

The y-distributions expected for ν and $\bar{\nu}$ may be
directly obtained on the basis of the present phenomenological theory of low energy weak interactions
and lepton-nucleon scattering theory. Assuming scale
invariance, one has

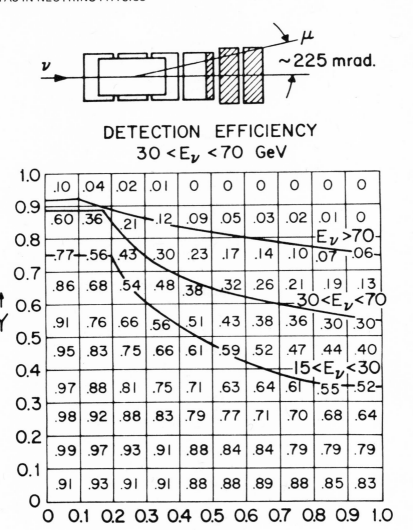

Figure 7 Detection efficienty ε_μ of the apparatus. The solid lines are geometric limits determined exclusively by the angular acceptance of the muon magnetic spectrometer, and by ranging out of muons with energies less than 4 GeV. The numerical values of ε_μ correspond to the (sample) energy interval $30 < E_{\nu,\bar\nu} < 70$ GeV.

$$\left(\frac{d\sigma}{dy}\right)^{\nu} = \frac{G^2 m E_{\nu}}{\pi} \int F_2(x)dx[1-y(1-B^{\nu}) +\frac{y^2}{2}(1-B^{\nu})+\frac{y^2}{2} R_L^{\nu}], \quad (1)$$

$$\left(\frac{d\sigma}{dy}\right)^{\bar{\nu}} = \frac{G^2 m E_{\bar{\nu}}}{\pi} \int \bar{F}_2(x)dx[1-y(1+B^{\bar{\nu}})+\frac{y^2}{2}(1+B^{\bar{\nu}})+\frac{y^2}{2} R_L^{\bar{\nu}}], \quad (2)$$

with

$$B^{\nu} = -\int x F_3(x)dx/\int F_2(x)dx, \quad B^{\bar{\nu}} = -\int x \bar{F}_3(x)dx/\int \bar{F}_2(x)dx,$$

and

$$R_L^{\nu} = \int [2 x F_1(x) - F_2(x)]dx/ \int F_2(x)dx,$$

$$R_L^{\bar{\nu}} = \int [2 x \bar{F}_1(x) - \bar{F}_2(x)]dx/ \int \bar{F}_2(x)dx,$$

where the $F_i(x)$ are dimensionless, scale-invariant nucleon structure functions.

If charge symmetry invariance is valid, the simplifying relations hold, viz,

$$F_2 = \bar{F}_2,$$
$$B^{\nu} = B^{\bar{\nu}},$$
$$B_L^{\nu} = R_L^{\bar{\nu}}. \quad (3)$$

and

Furthermore, if parity is maximally violated, i.e., if the space-time structure of the hadronic weak current is pure V-A,

$$B^{\nu} = B^{\bar{\nu}} = 1, \quad (4)$$

and we obtain

$$(d\sigma/dy)^{\nu} = \frac{G^2 m E_{\nu}}{\pi} \int F_2(x)dx,$$
$$(d\sigma/dy)^{\bar{\nu}} = \frac{G^2 m E_{\nu}}{\pi} (1-y)^2 \int F_2(x)dx, \quad (5)$$

$$\frac{\sigma^{\bar{\nu}}}{\sigma^{\nu}} = \frac{2-B}{2+B} \approx \frac{1}{3} \, ,$$

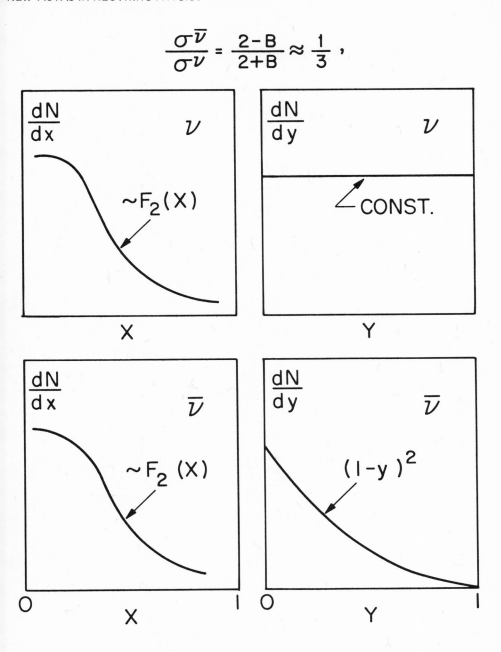

Figure 8   Distributions in the Bjorken variables x and
            y expected from simplified theory (see text).

and

$$\sigma^{\bar{\nu}}/\sigma^{\nu} = 1/3. \tag{6}$$

These simplified distributions by y are shown in Fig. 8.

The observed distribution in y for ν and $\bar{\nu}$ for $10 < E_{\nu,\bar{\nu}} < 30$ GeV are shown in Fig. 9. Note how the detection efficiency has sagged above y = 0.6, which indicates the incompleteness of the model-independent efficiency correction above that y-value in this energy region (Fig. 7). Fitting the data below y = 0.6 in Fig. 9, however, yields $B^{\nu} = 0.60 \pm 0.30$ and $B^{\bar{\nu}} = 0.94 \pm 0.09$, in approximate agreement with the simplified theoretical distributions shown in Fig. 8, if R_L^{ν} and $R_L^{\bar{\nu}}$ are taken to be zero.

In Fig. 10 are shown the y^{ν}- and $y^{\bar{\nu}}$-distributions in the energy region $E_{\nu,\bar{\nu}} > 70$ GeV, with average energies $<E_{\nu}> = 126$ GeV and $<E_{\bar{\nu}}> = 106$ GeV. Here the region of good detection efficiency extends to y = 0.85. Out to that limit the y^{ν}-distribution is fit with $B^{\nu} = 0.83 \pm 0.20$, again consistent with theoretical expectations, and indicating that the almost uniform y-distribution for ν is insensitive to the precise value of B^{ν}. In contrast, the $y^{\bar{\nu}}$-distribution in Fig. 10 is best fit with $B^{\bar{\nu}} = 0.41 \pm 0.13$, roughly 4 standard deviations lower than the value of $B^{\bar{\nu}}$ determined in the lower energy interval, $10 < E_{\bar{\nu}} < 30$ GeV, in Fig. 9. This is the most direct, though perhaps not the clearest, manifestation in the data of the high-y anomaly: the form of the y^{ν}-distribution remains constant with changing ν energy while the form of the $y^{\bar{\nu}}$-distribution changes with increasing $\bar{\nu}$ energy.

The complete energy dependence of the y-distributions

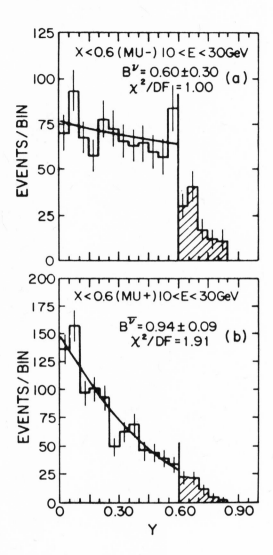

Figure 9   Distributions in y for (a) 946ν events and
(b) 991 $\bar{\nu}$ events with 10<$E_{\nu,\bar{\nu}}$<30 GeV. The
cross hatched areas are in a y-region in which
the correction for detection efficiency is incomplete (Fig. 7); they are not included in the
fits for $B^{\nu,\bar{\nu}}$.

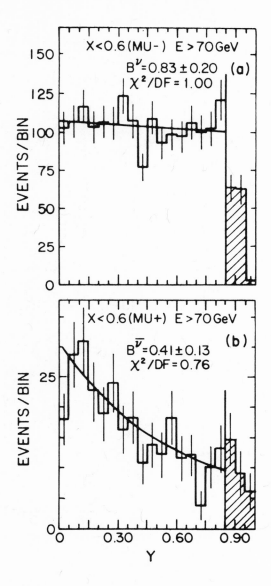

Figure 10   Distributions in y for (a) 1905 ν events with
            $<E_ν>$ = 126 GeV, and (b) 310 $\bar{ν}$ events with
            $<E_{\bar{ν}}>$ = 106 GeV.   All events have $E_{ν,\bar{ν}}$ > 70 GeV.
            The cross hatched areas mean the same as in
            Fig. 9, but note the improvement in detection
            efficiency with energy.

is shown in Fig. 11 where we plot the first moments
$<y^{\nu}>$ and $<y^{\bar{\nu}}>$ (for all x < 0.6) as functions of E_{ν}
and $E_{\bar{\nu}}$ up to 150 GeV. The $<y^{\nu}>$ data are not constant
at $<y^{\nu}> = 0.5$, as might be expected, because of the
regions of incomplete detection efficiency, i.e., loss
of events at large-y (Fig. 7), and also to a small
extent because of the cut at x = 0.6. That loss has
been taken into account in the Monte Carlo calculation
of $<y^{\nu}>$ with which data in Fig. 11 are compared. Again,
note that the measurements of $<y^{\nu}>$ do not determine a
precise value of B^{ν}, but are consistent within experimental error with a large value of B^{ν}. On the other
hand, the data for $<y^{\bar{\nu}}>$, which correspond to $B^{\bar{\nu}} = 0.9$
at $E_{\bar{\nu}} < 30$ GeV, rise sharply to a significantly higher
value of $<y^{\bar{\nu}}>$, and therefore to a significantly lower
value of $B^{\bar{\nu}}$, at higher $E_{\bar{\nu}}$. There is, however, no
single value of $B^{\bar{\nu}}$ which fits the data over the entire
energy range 10 to 150 GeV. This is shown by comparison
with the Monte Carlo calculated curves for two values
of $B^{\bar{\nu}}$, which are based on the same detection efficiency
corrections used to obtain the calculated curves for
$<y^{\nu}>$. Fig. 11 exhibits most clearly the high-y anomaly,
i.e., the abrupt change with $E_{\bar{\nu}}$ of the form of the $y^{\bar{\nu}}$-
distribution. We are not aware of any energy dependent
misbehavior of the experimental apparatus that might
introduce a spurious, apparent energy threshold in
$<y^{\bar{\nu}}>$ in the vicinity of 30 GeV, and which would lead
at higher energies to such large values of $<y^{\bar{\nu}}>$.

Within the context of the quark-parton model with
three quarks (u, d and s), it is possible to evaluate
the fraction of antiquarks indicated by the $y^{\bar{\nu}}$-dis-
tributions below 30 GeV and well above that energy.
The parameter $B^{\bar{\nu}}$ in eq. (2) is related to the fraction

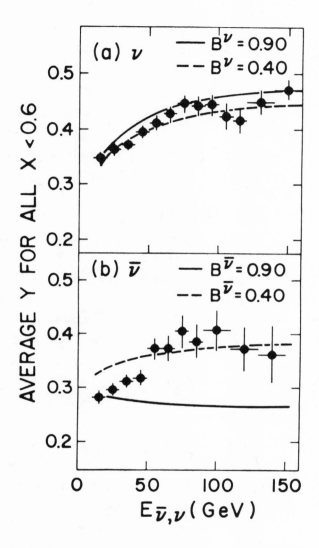

Figure 11   First moments of the y-distributions <u>vs</u>
energy for (a) ν events and (b) $\bar{\nu}$ events.

of antiquarks by the equation

$$B^{\bar{\nu}} = 1 - 2 \int \bar{q}(x)dx / \int [\bar{q}(x) + q(x)]dx, \qquad (7)$$

where $q(x)$ and $\bar{q}(x)$ are the momenta in the non-strange quarks and antiquarks. Thus, the curves marked $B^{\bar{\nu}} = 0.90$ and $B^{\bar{\nu}} = 0.40$ in Fig. 11 correspond to $\int \bar{q}(x)dx / \int [\bar{q}(x) + q(x)]dx = 0.05$ and 0.30, respectively, but recall also that the data for $\langle y^{\bar{\nu}} \rangle$ are not fit over the entire energy region by a single value of $B^{\bar{\nu}}$.

From Fig. 9 the measured $y^{\bar{\nu}}$-distribution in the energy interval $10 < E_{\nu} < 30$ GeV yields for the ratio $\int \bar{q}(x)dx / \int [\bar{q}(x) + q(x)]dx = 0.05 \pm 0.05$. This is consistent with the value 0.09 ± 0.04 obtained from data[13] at $E_{\nu,\bar{\nu}} < 10$ GeV, and with the limit of 0.10 from recent muon-nucleon scattering data.[14] Consequently, $\bar{\nu}$ scattering from the antiquarks of the conventional three-quark model is too small to account for the magnitude of the high-y anomaly. Furthermore, the model is unable to explain the energy threshold for the high-y anomaly.

If new hadron production is indeed the source of the high-y anomaly and dimuons, it is of interest to consider the dependence of the ν and $\bar{\nu}$ charged current total cross sections $\sigma_c^{\nu,\bar{\nu}}$ on energy. One or both of these might be expected to show a threshold effect if the cross section measurements are sufficiently accurate. experimentally, it is more direct and more accurate to study the ratio of cross sections $\sigma_c^{\bar{\nu}}/\sigma_c^{\nu}$, because most systematic apparatus effects cancel out of the ratio, and it is possible to make a relatively precise measurement of the ratio that is independent of the ν

and $\bar{\nu}$ fluxes.

We show in Fig. 12 the measured ratio $\sigma_c^{\bar{\nu}}/\sigma_c^{\nu}$ in the energy region from 10 to 60 GeV. Here the relative normalization[15] of $\sigma_c^{\bar{\nu}}$ to σ_c^{ν} is provided by ν and $\bar{\nu}$ events involving quasielastic scattering and N^*-pro- duction, which do in fact appear clearly in invariant mass (W) plots of the data. The signature of these events is taken to be the joint requirement $W < 1.6$ GeV and $Q^2 < 1.5$ GeV^2. Above 60 GeV the fraction of all events satisfying this requirement diminishes, which accounts for the absence of points above 60 GeV in Fig. 12.

The justification for using quasielastic scattering and resonance production for normalization lies in the fact that, for $E_{\nu,\bar{\nu}} \gtrsim 3$ GeV,

$$\sigma(\nu_\mu + N \rightarrow \mu^- + N \text{ or } N^*) = \sigma(\bar{\nu}_\mu + N \rightarrow \mu^+ + N \text{ or } N^*), \tag{8}$$

which is known empirically and also expected on very general theoretical grounds. An alternative normal- ization procedure of greater generality arises from the relation

$$E_\nu^{\lim_{\nu \to \infty}} d\sigma/dW^2 \left[\nu_\mu + \binom{T=0}{\text{target}}\right) \rightarrow \mu^- + \text{hadrons}]$$

$$= \lim_{E_{\bar{\nu}} \to \infty} d\sigma/dW^2 [\bar{\nu}_\mu + \left(\begin{matrix} T=0 \\ \text{target} \end{matrix}\right) \rightarrow \mu^+ + \text{hadrons}], \tag{9}$$

initially obtained by Lee and Yang[16] and others[17], who recognized that

$$\lim_{E \to \infty} d\sigma/dW^2 dq^2 = \text{const } W_2(q^2, W^2), \tag{10}$$

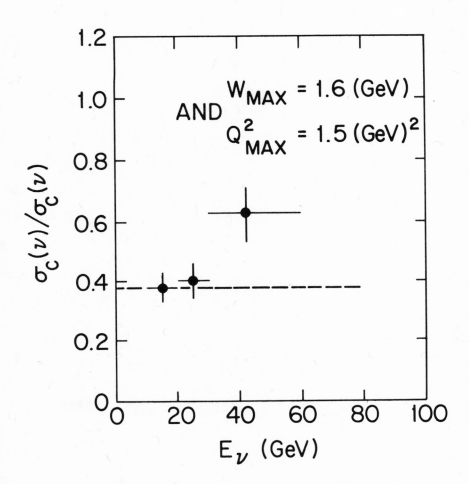

Figure 12 Ratio of antineutrino to neutrino charged
current cross sections obtained by the
quasielestic flux normalization method.

which, in conjunction with charge symmetry invariance,
leads to eq. (9). The validity of eq. (9) at finite
values of $E_{\nu, \bar{\nu}}$ has been discussed by Sakurai[18] who
specifies the upper limits of the integrals for
which

$$\int_0^{W_{MAX}} \frac{d\sigma_c^{\bar{\nu}}}{dW^2} \, dW^2 = \int \frac{d\sigma_c^{\nu}}{dW^2} \, dW^2 \tag{11}$$

in a given energy region with a given precision. The
range of energy over which $\sigma_c^{\bar{\nu}}/\sigma_c^{\nu}$ may be obtained from
the data is extended by this normalization procedure,
as is shown in Fig. 13 in which we have utilized the
Sakurai prescription and variations thereof.

     To obtain the experimental points in Figs. 12 and
13, it is necessary to compensate for the incomplete
detection efficiency correction described above in a
model dependent way. The magnitude of this compensation
is moderate; for example, above 50 GeV, it is less
than 20% for ν and less than 2% for $\bar{\nu}$. Note that the
results obtained by the two normalization methods are
in agreement. Below 30 GeV, one obtains $\sigma_c^{\bar{\nu}}/\sigma_c^{\nu} = 0.38$
± 0.06, consistent with the value expected from the
simplified theory [eq. (6)], but above that energy
the ratio is observed to rise significantly above that
value.

     The picture that emerges from these data is strongly
supportive of the hypothesis of new hadron production.
The ν and $\bar{\nu}$ single muon data, taken together, show
effective violations of scale invariance and charge
symmetry invariance which are explained by that
hypothesis. Dimuons are produced by both ν and $\bar{\nu}$, but
apart from dimuon production by ν, there is no direct
manifestation of new particle production by ν in the

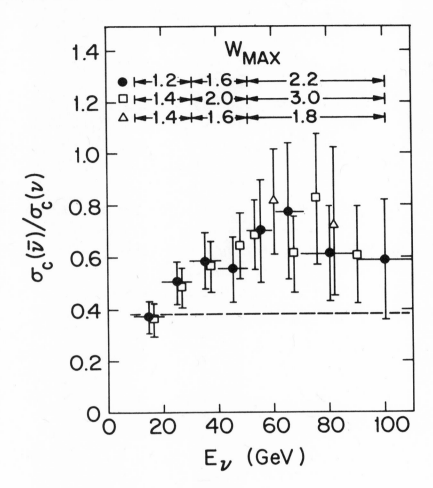

Figure 13   Determination of the ratio of antineutrino to neutrino charged current cross sections by the Sakurai flux independent normalization prescription (black dots). W_{max} refers to the maximum hadronic recoil mass (W) that is used in the normalization procedure. The open triangle, circle and square denote the value of the ratio that is obtained if W_{max} is varied as shown in the given energy intervals.

(single muon) data. It can, however, be argued that
new hadron production by ν, if present at the level
of 10% or even 20% of σ_c^ν, would be quite difficult
to observe in either y^ν- or W^ν-distributions. It
appears from the high-y anomaly and the behavior of
$\sigma_c^{\bar\nu}/\sigma_c^\nu$ that the effect is larger for $\bar\nu$ and therefore
more easily observed; also the form of $y^{\bar\nu}$-distribution
particularly lends itself to a search for anomalies
in $\bar\nu$-interactions.

The material of this talk is the product of the
Harvard-Pennsylvania-Wisconsin-Fermilab collaboration
which includes the following individuals: A. Benvenuti,
D. Cline, W. T. Ford, R. Imlay, T. Y. Ling, A. K.
Mann, D. D. Reeder, C. Rubbia, R. Stefanski, L. Sulak
and P. Wanderer.

REFERENCES

1. A. Benvenuti et al., Phys. Rev. Lett. $\underline{33}$, 984 (1974).

2. Ibid., $\underline{34}$, 597 (1975).

3. Ibid., $\underline{34}$, 419 (1975).

4. Ibid., $\underline{35}$, 1199 and 1203 (1975).

5. Ibid., $\underline{35}$, 1249 (1975).

6. See, for example, L. N. Chang, E. Derman and J. Ng, Phys. Rev. Lett. $\underline{35}$, 6 (1975).

7. A. Pais and S. B. Treiman, Phys. Rev. Lett. $\underline{35}$, 1206 (1975).

8. Preprint HPWF-76/1, "Further Data on the High-y Anomaly in Inelastic Antineutrino Scattering", A. Benvenuti et al., submitted to Phys. Rev. Lett.

9. Preprint HPWF-76/2, "A Measurement of the Ratio $\sigma(\bar{\nu}_\mu + N \rightarrow \mu^+ + X) / \sigma_c(\nu_\mu + N \rightarrow \mu^- + X)$ at High Energy", A. Benvenuti et al., submitted to Phys. Rev. Lett.

10. A. Benvenuti et al., Colloq. Int. CNRS $\underline{245}$, 397 (1975).

11. A. Benvenuti et al., NIM $\underline{125}$, 447 (1975).

12. Ibid., $\underline{125}$, 457 (1975).

13. H. Deden et al., Nucl. Phys. $\underline{B85}$, 269 (1975).

14. L. Mo, Int'l Conf. on Lepton and Photon Interactions at High Energy, Stanford University, Aug. 1975.

15. A. Benvenuti et al., Phys. Rev. Lett. $\underline{32}$, 125 (1974).

16. T. O. Lee and C. N. Yang, Phys. Rev. Lett. $\underline{4}$, 307 (1960), Phys. Rev. $\underline{126}$, 2239 (1962).

17. Y. Yamaguchi, Prog. Theor. Phys. $\underline{23}$, 1117 (1960); N. Cabibbo and R. Gatto, Nuovo Cimento $\underline{15}$, 305 (1960); S. L. Adler, Phys. Rev. 143, 1144 (1966).

18. J. J. Sakurai, "Flux Determination in Neutrino

Experiments at NAL Energies", preprint UCLA/73/TEP/7
1973 (unpublished).

SOME RECENT WORK ON ZERO MASS FIELD THEORIES[*]

Thomas Appelquist[†]

Department of Physics

Yale University, New Haven, Connecticut   06520

ABSTRACT

Some new results on the zero mass behavior of
renormalizable field theories are reviewed. Much of
the work is directed toward the problem of defining
finite transition probabilities in theories with coupled
massless fields. In particular, it has been shown
through lowest non-trivial order that the production,
scattering and detection (with finite energy resolution)
of massive quarks in the unbroken SU(3) color gauge model
is free of infrared divergences in perturbation theory.

I. INTRODUCTION

The study of Feynman amplitudes as some (or all) of
the mass parameters vanish has played an important role
in the development of quantum field theory. The infrared
divergence problem in quantum electrodynamics[1,2] is

---
[*]Research (Yale Report COO-3075-135) supported in part
by U.S.E.R.D.A. under contract no. AT(11-1)3075.
[†]Alfred P. Sloan Foundation Fellow.

perhaps the first example and much of the modern work
on the scale invariance and the renormalization group
is based on the analysis of mass singularities in
pertrubation theory[3,4]. Large momentum behavior in
renormalizable field theory is very intimately connected to zero mass behavior. The discovery of asymptotic
freedom[5] and the emergence of the colored quark-gluon
model[6] as a possible strong interaction theory has
stimulated even more interest in this problem. The
most fundamental issue here is the nature of the quark
confinement mechanism and the true asymptotic states.

     During the last few months, several papers have
appeared dealing with mass singularity problems. They
are mainly technical extensions of earlier work. In
this lecture I will start by reminding you of some of
the classic results and then list with some commentary
the ongoing work that I am aware of. I will omit the
work of Cornwall and Tiktopolous[7] which will be described
in this session by Professor Cornwall. In the last part
of the talk, I will describe in more detail some work
in which I was involved[8]. This work indicates that in
theories with coupled massless fields (such as the
Yang-Mills quark gluon theory), certain partial transition probabilities are free of infrared divergences.
In particular, the production, scattering and color-blind
detection of massive quarks with finite energy resolution
detectors is finite order by order in an appropriately
defined coupling constant. This result is proven to
lowest non-trivial order and we conjecture that it is
true to any order. After outlining the result, I will
discuss its implications.

II. HISTORY

It is useful to consider a specific simple process
to remind you of the old results. The same process
and the same notation will be used to discuss the newer
work. Consider first, in ordinary quantum electrodynamics, the production of electrons and photons by
the external local current $j_\mu(x) = \bar{\psi}_e(x)\, \gamma_\mu \psi_e(x)$.
Suppose that the electron is detected, as shown in
Fig. 1, with some finite energy resolution ΔE.

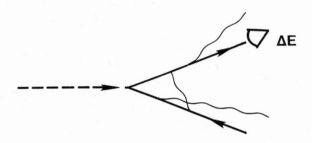

Fig. 1. Production by an external current in
quantum electrodynamics.

Using a photon mass λ to regulate infrared divergences,
we can define a dimensionless transition probability by
dividing the corresponding cross section by the Born
cross section. This gives $R_{\Delta E}(E/m_e,\ \Delta E/m_e,\ \lambda/m_e,\ \alpha)$,
where E represents the energies in the problem (center
of mass, detected electron) and angular dependence
is suppressed. The well known result[1,2] is that the
limit $\lambda \to 0$ exists to any order in α by virtue of cancellations between virtual photon exchange and real
photon emission.

When a theory contains coupled massless fields (as
does the Yang-Mills theory), the infrared behavior
becomes considerably more complicated. The limit $m_e \to 0$

in quantum electrodynamics illustrates the problem. If
an on-shell photon of momentum $k_\mu(k^2=0)$ is emitted from
an on-shell electron of momentum $p_\mu(p^2=0)$, logarithmic
infrared divergences arise both from $\vec{k}\to0$ and $\vec{p}\to0$ in the
phase space integral. In addition, as $\vec{k}$ becomes
parallel to $\vec{p}$, a logarithmic divergence arises from the
singular parent electron propagator

$$\frac{1}{(k+p)^2} = \frac{1}{2k\cdot p} = \frac{1}{2|\vec{k}||\vec{p}|(1-\cos\theta)} \quad . \qquad (1)$$

All these divergences pile on top of one another, creating
a cloud of logarithmic infinities. Nevertheless, it
was shown by Kinoshita[3] and Lee and Nauenberg[4] that in
certain total transition probabilities, even these
divergences cancel.

In the case of production by the external current
$j_\mu(x) = \bar\psi_e(x)\,\gamma_\mu\,\psi(x)$, it is the total cross section
that is finite. This is proven by first showing that
each graph contributing to

$$\Pi_{\mu\nu}(q) = \int d^4q\ e^{iq\cdot x}<0|T\ j_\mu(x)\ j_\nu(0)|0>$$

is finite in the limit $m_e\to0$. This follows from a power
counting analysis valid for $q^2>0$ as well as $q^2<0$. The
proof of Kinoshita[3] extends to all orders and covers
any renormalizable field theory except the Yang-Mills
theory. It then follows that the sum of the contributions to the total cross section corresponding to
the various cuts of a given graph in $\Pi_{\mu\nu}(q)$ will be
finite as $m_e\to0$. The only condition on this result is
that infrared divergences must not be introduced through
charge renormalization. This can be insured by

subtraction at a Euclidean point $q^2 = -M^2$, leading
to a corresponding coupling constant $e(M)$. The total
cross section, divided by the Born cross section, is
$R(E/M, m_e/M, \lambda/M, e(M))$ and the limit $m_e \to 0$ and $\lambda \to 0$ is
finite order by order in $e(M)$.

The finiteness of total rates even when $m_e \to 0$ was
originally pointed out in the context of μ meson decay
with the radiative corrections computed to order α[9]. It
was also noticed there that certain partial rates,
corresponding to the electron and radiated photon
moving in nearly the same direction, also remained
finite. This observation has been stated more precisely and proven to all orders in a recent paper by
Sterman[10]. I'll say a little more about this work
in a moment.

III. SOME CURRENT WORK

The work that I'm aware of falls into three
categories:

1. Generalizations of the Kinoshita analysis of
   Green's functions either to include Yang-Mills
   theories[11] or to include more general momentum
   configurations[12].

2. The construction of finite partial transition
   rates for massless theories[10]. This work has
   not yet dealt with Yang-Mills theories.

3. The specific problem of production and scattering
   in the Yang-Mills theory[8].

Let me say a little more about the first two and then
turn to a discussion of number three in the next section.

1. Kinoshita and Ukawa[12] are engaged in a rather
general and systematic study of the zero mass properties
of Feynman amplitudes. They are extending the old

analysis to include on-shell and exceptional momentum
configurations. This is important ground work for
many problems of physical interest.

Poggio and Quinn[11] have recently completed an
extension of the old Kinoshita result to include
Yang-Mills theories. They have shown that Euclidean
Green's functions are free of mass singularities and
therefore, by analytic continuation, that certain total
production cross section are also finite. For example,
the total cross section for production of quarks and
gluons by an external electromagnetic current is finite
to any finite order in the coupling constant. This had
already been checked in low orders and had been assumed
to all orders in renormalization group treatments of
e^+e^- annihilation.

2. Sterman[10] has shown how to precisely define
infrared finite <u>partial</u> transition rates even in completely massless field theories. In the case of
production by an external local current, these correspond
to jet-like configurations as shown in Fig. 2.

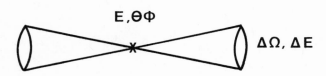

E,ΘΦ

ΔΩ, ΔE

Fig. 2 Center of Mass Jets for Production
by an External Current.

E is the total energy and θ, φ describe the jet axis
orientation. ΔΩ is the angular aperture of the cone
and ΔE is the energy of soft quanta which leak out of

the cone. The finiteness has been established to all
orders. This result, if extended to asymptotically
free theories, could provide a foundation for a
renormalization group treatment of jet structure in
e^+e^- annihilation.

IV. INFRARED FINITENESS IN YANG-MILLS THEORIES

The work I will describe applies to field theories
with massive fields along with coupled massless fields[8].
The quark gluon theory with massive quarks and an unbroken SU(3) color gauge symmetry is an example of
such a field theory. My discussion will refer to that
theory, with a single quark flavor for simplicity,
but it really applies to any such field theory.

The question is whether the production scattering
and detection of single quarks with energy resolution
ΔE is finite in perturbation theory. Here one is asking
a more detailed question than whether total transition
rates or transition rates corresponding jet structures
are finite. It is more like the conventional infrared
problem in ordinary quantum electrodynamics but now
complicated by the angular divergences that appear in
any theory with coupled massless fields (Eq. 1). It is
very closely connected to the question of the existence
of asymptotic states with quark quantum numbers.

Since we are discussing asymptotic quarks, it is
appropriate to define the renormalized quark mass by
subtraction on mass shell, $p^2 = \mu^2$. The coupling constant
g(M) is defined by subtracting at some Euclidean scale
M. As an example I will again consider production by an
external local current—this time the color singlet
current $J_\mu(x) = \sum_i \bar{q}_i(x) \gamma_\mu q_i(x)$. A detector is then

arranged (Fig. 3) to trigger on a single quark of
momentum $\bar{p}$ and energy $E = +\sqrt{p^2+\mu^2}$ with some resolution
ΔE. The detector is color-blind so that, for example,
gluons do not register. The corresponding transition
probability is

$$R_{\Delta E}(E/M,\ \Delta E/M,\ \mu/M,\ \lambda/M,\ g(M)),$$

where λ represents some infrared cutoff such as a gluon
mass or a dimensional parameter associated with dimensional continuation[13].

The computation of $R_{\Delta E}$ is, as before, organized
according to the graphs in

$$\Pi_{\mu\nu}(q) = \int e^{iq\cdot x}\ d^4x\ <0|T\ J_\mu(x)\ J_\nu(0)\ |0>\ .$$

$R_{\Delta E}$ gets a contribution from each cut of each graph but
since the quark phase space is restricted by the detector kinematics, the cuts are not full unitarity cuts.
Since the detector is color blind, the group theory
traces can be done for a given graph before considering
each cut. Only electrodynamic-like graphs enter through
order $g^2(M)$ and therefore $R_{\Delta E}$ is cearly finite to this
order as $\lambda \to 0$. In next order (3 loops in $\Pi_{\mu\nu}(q)$), the
gluon self couplings enter. We have proven that the sum
of the contributions to $R_{\Delta E}$ corresponding to each of
the cuts of a any graph of this order $\Pi_{\mu\nu}(q)$ is finite
as $\lambda \to 0$[8]. The computation is quite lengthy for the graphs
of Fig. 3.

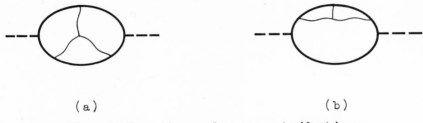

(a)   (b)

Fig. 3 Two three-loop contributions
to $\Pi_{\mu\nu}(q)$.

We conjecture that $R_{\Delta E}(E/M, \Delta E/M, \mu/M, 0, g(M))$ is
finite to all orders in $g(M)$. As a final comment, I
should point out that graphs with Fadeev-Popov ghost
loops enter $\Pi_{\mu\nu}(q)$ at this and higher levels. Some
Cutkosky cuts go through them and the ghosts then play
the usual role of canceling unphysical longitudinal
gluons.

I will conclude with some commentary about this
result.

1. The computations have been done in Feynman
gauge. It would be worthwhile to repeat the analysis
in a general gauge and directly check the gauge
invariance of the finiteness.

2. A proof to all orders is not trivial but
probably possible. No essentially new features enter
in higher orders.

3. Scattering problems can be dealt with in the
same way. We have shown that quark scattering by an
external color singlet field followed by color blind
detection is finite to order $g^4(M)$ (two loop corrections
to the elastic amplitude and up to two undetected gluons

in the final state). Quark-quark scattering has been
shown to be finite by Yao[14] at the single gluon
emission level.

4. The result is perturbative and independent
of renormalization group considerations. It is true
for the asymptotically free theory discussed here as
well as non-asymptotically free models[8]. However,
the utility of perturbation theory for computing $R_{\Delta E}$
is certainly renormalization group dependent. In the
color gauge theory, perturbation theory could be useful
with E, ΔE, and μ taken large enough. However, the
behavior as $\Delta E \rightarrow 0$ (a well known result in quantum
electrodynamics) is a strong coupling problem in this
theory.

5. The essential difference between this result
and the corresponding one in quantum electrodynamics
is that here $R_{\Delta E}$ is not expressed in terms of an on-shell
coupling constant. As a consequence, there is no immediate
classical correspondence.

6. To restate the result, there is no sign in
perturbation theory of the impossibility of the prod-
cution of asymptotic states with quark quantum numbers
(e.g. fractional electric charge). This result requires
at least some comment on the confinement problem.

My guess is that to see confinement, the perturbation
expansion would have to be reorganized into something
like a skeletor expansion. The mechanism is probably
indicated by the two dimensional Yang-Mills model[15].
There the bare gluon propagator $1/q^2$ corresponds to a
linearly growing potential in coordinate space which
permanently confines the quarks. If one looks at the
production of quarks by a group-singlet current, the

confinement is signaled in perturbation theory by the
presence of infrared divergences which then conspire
to all orders to make the production probability vanish.
This feature might be seen in the skeleton expansion of
the four dimensional theory. If the two, three and four
point functions in each skeleton are appropriately enhanced in the infrared (from a strong-coupling solution
to the renormalization group equations), divergences
would appear in each skeleton. A summation over all
skeletons might then lead to a zero production
probability for states with quark quantum numbers.

REFERENCES

1. D. R. Yennie, S. C. Frautschi and H. Suura, Ann.
    Phys. (N.Y.) 13, 379 (1961), and references cited
    therein.

2. Work on the coherent state approach can be traced
    through the recent papers of D. Zwanziger and
    references cited therein. D. Zwanziger, Phys. Rev.
    Lett. 30, 934 (1973) and Phys. Rev. D7, 1082 (1973).

3. T. Kinoshita, J. Math. Phys. 3, 650 (1962).

4. T. D. Lee and M. Nauenberg, Phys. Rev. 133B, 1549
    (1964).

5. H. D. Politzer, Phys. Rev. Lett. 26, 1346 (1973).
    D. J. Gross and F. Wilczek, Phys. Rev. Lett. 26,
    1343 (1973).

6. H. Fritzsch, M. Gell-Mann and H. Leutwyler, Phys.
    Lett. 47B, 365 (1973).

7. J. M. Cornwall and G. Tiktopolous, Phys. Rev. Lett.
    35, 338 (1975) and UCLA preprint UCLA/75/TEP/21,
    October 1975.

8. T. Appelquist, J. Carazzone, H. Kluberg-Stern and
    M. Roth, Yale University Preprint, January 1976.

9. This was, in fact, suggested earlier by T. Kinoshita
    and A. Sirlin, Phys. Rev. 113, 1652 (1959).

10. G. Sterman, University of Illinois preprint, December
    1975. Some discussion about the finiteness of
    partial rates can also be found in references 3 and 4

11. E. Poggio and H. Quinn, Harvard University Preprint,
    February 1975.

12. T. Kinoshita and A. Ukawa, Cornell University Preprint
    CLNS-322, November 1975.

13. The use of dimensional continuation for infrared re-
    gulation has been discussed by several authors.

REFERENCES

See, for example, R. Gastmans, J. Verwaest and
R. Meuldermans, University of Leuvain Preprint,
November 1975 and references therein.

14. Y. P. Yao, University of Michigan Preprint, December
     1975.

15. G. 't Hooft, Nucl. Phys. <u>B75</u>, 461 (1974), C. G.
     Callan Jr., N. Coote and D. J. Gross, Princeton
     University Preprint, October 1975.

INFRARED SINGULARITIES OF NON-ABELIAN GAUGE THEORIES*

John M. Cornwall and George Tiktopoulos

(Presented by John M. Cornwall)

University of California

Los Angeles, California   90024

For some time, we have been interested in the infrared structure of non-Abelian gauge theories.[1,2]
There are prominent infrared singularities near the mass
shell only when the local gauge invariance is unbroken,
as in quantum chromodynamics (QCD). We (and other
authors[3,4]) began with the straightforward calculation
of Feynman graphs, in leading-logarithm approximation,
regulating the infrared singularities either by staying
slightly off the mass shell by an amount characterized
by a small mass μ, or by introducing a fictitious (group-
symmetric) vector mass μ in the Feynman gauge. The results turned out to be much simpler than the complexity
of the calculations suggested, and are characterized by
certain properties of exponentiation and factorization
which are reminiscent of the corresponding results for
the Abelian case (quantum electrodynamics, or QED).
These properties are summarized by a differential

---

*Work supported in part by the National Science Foundation.

equation in μ (the infrared cutoff) which looks very
much like a renormalization group (RG) equation with a
special type of anomalous dimension generated by the
near-mass-shell infrared singularities. In fact, this
differential equation is considerably easier to understand than a mere catalog of the graphical experiments,
and we consider its further development essential to the
explication of such obviously non-perturbative problems
as confinement in QCD. It appears that this equation,
born in perturbation theory, can transcend its humble
origins and yield a signal for confinement even at the
simplest level of application. As yet, we do not have a
complete derivation of the equation to all orders for
non-Abelian gauge theories, but it can be proved readily
for QED. In all cases known to us, the results of the
equation are consistent with graphs calculated up to
sixth and sometimes eighth order.

1.   THE μ ∂/∂μ EQUATIONS

There is not one, but many differential equations
appropriate to the study of infrared singularities of
gauge theories. These equations differ in detail
according to the nature of the infrared cutoff μ (vec-
tor-meson mass or alternatively distance from the mass
shell; this latter case has an interesting space-time
interpretation) and according to the kinematic regime
considered. Common to these regimes is that they are
all on (or near) the mass shell. The differential
equations all have the general structure (for <u>leading</u>
infrared singularities)

$$\mu \frac{\partial}{\partial \mu} T_r = \sum_{ij,s} \Gamma_{rs}^{(ij)} T_s \quad , \tag{1}$$

where T_r is an on-shell (or nearly on-shell) amplitude,[5] with external momenta p_i ($p_i^2 = M_i^2$ on shell). The index r characterizes the various color channels. The $\Gamma_{rs}^{(ij)}$ are matrices which may depend on μ and on the invariants $p_i p_j$. The sum over i,j runs over pairs of external colored on-shell lines, corresponding to the ways of joining such lines by gluons. In certain cases, the product ΓT in (1) is replaced by a convolution in momentum space. The rule for forming the differential equation (1) is easily stated; we give it only for the case when μ is a gluon-mass cutoff: Join all pairs of external legs on the truncated on-shell amplitude T by gluon lines, supplying free-particle propagators for the external legs. The external-leg bare vertices are constructed in the usual way. For each such gluon line there is a factor $(\mu \, \partial/\partial\mu)\Delta$, where Δ is the gluon propagator. Integrate over the momenta of these gluons, keeping T on-shell (that is, p_i^2 is strictly equal to M_i^2, for all external legs i). However, if some momentum transfer $p_i - p_j$ has components $O(\mu)$, the dependence of T on the gluon momenta via this momentum transfer must be retained. Details of the derivation of this rule are given in Ref. 2. It is based on a low-energy formula for the emission of soft gluons whose form (but not whose derivation!) looks like the leading term of the Low theorem.

The three main kinematic regimes are: (1) the fixed-angle regime, in which all non-trivial invariants $p_i p_j$ ($i \neq j$) are large compared to the fermion masses the M_i^2 as well as to μ^2. This is the simplest regime, since $\Gamma_{rs}^{(ij)}$ has the form $\Gamma^{(i)}\delta_{ij}\delta_{rs}$. Then there is (2) the infrared regime, where the $p_i p_j$ need not be large compared to M_i^2 (if i is not a gluon line), but μ^2 is small compared to masses and invariants. Finally we have (3) the near-forward regime, an example of which is quark-quark elastic scattering with $s, M^2 >> |t| >> \mu^2$. In this regime the momentum-space convolution is relevant, a we discuss later.

In the fixed-angle regime, the leading-logarithm $\Gamma^{(i)}$ are given by

$$\Gamma^{(i)} = \frac{g^2}{8\pi^2} C_2(i) \, \ln(-t/\mu^2) \, , \qquad (2)$$

where t is a typical large invariant, and $C_2(i)$ the quadratic Casimir eigenvalue appropriate to the ith leg. The solution to (1) is a product of form factors, each of which exponentiates, times the Born term.

In the infrared regime, the form factor F_R for the process $R\bar{R} \to$ group-singlet current, where R denotes a massive particle in representation R of SU(3) of color, obeys

$$\mu \frac{\partial}{\partial \mu} F_R = \frac{g^2}{4\pi^2} C_2(R) \, H(t/M^2) \, F_R, \qquad (3a)$$

$$F_R \sim \exp\left[\frac{g^2}{8\pi^2} C_2(R) \, H(t/M^2) \, \ln\mu^2\right], \qquad (3b)$$

with

$$H(t/M^2) = \frac{1}{2}(2M^2 - t) \int_0^1 \frac{d\beta}{M^2 - \beta(1-\beta)t} \, . \qquad (4)$$

We note, for future reference, that the form factor for $R\bar{R} \to$ gluon, denoted $G_R(t)$, obeys (3) with $C_2(R) \to C_2(R) - \frac{1}{2}C_2(A$ where A refers to the adjoint (color octet) representation to which the gluons belong. Scattering amplitudes

obey more complicated equations. The essential feature
to note is that the emission of <u>real</u> gluons from any
process leads to additional damping for small μ, quite
unlike QED where the uncharged photons yield infrared
singularities only through phase-space divergences. Thus
the amplitude $T_{(N)}$ for elastic quark-quark scattering
with emission of N real gluons of momenta k_j obeys an
equation something like

$$\mu \frac{\partial}{\partial \mu} T_{(N)} = \left[\sum_j C_2(A) \frac{g^2}{8\pi^2} \ln \frac{p \cdot k_j}{M\mu} + \dots \right] T_{(N)}, \tag{5}$$

where p symbolizes a typical quark momentum, and the
omitted terms reflect the infrared singularities associated with $T_{(0)}$, the amplitude for no gluon emission.
We shall soon return to (5) and its important signal for
confinement. Other implications for hadronic physics
are discussed in Section 4. Perhaps the most important
is the absence of leading infrared logarithms for processes involving only colorless hadrons, considered as
clusters of free, colored particles.

We have been brief in discussing the fixed-angle
and the infrared regimes, since considerably more detail
can be found in Ref. 2. There are some new results for
the near-forward regime, which are described below at
greater length.

2. APPLICATIONS TO THE NEAR-FORWARD REGIME

By virtue of much recent work on summing graphs in
the near-forward regime[6] (some of it reported by McCoy
and Wu at this conference), it is now possible to com-
pare the near-forward solution of the $\mu \frac{\partial}{\partial \mu}$ equation to
"experiment." To eighth order the graphologists find

that the gluon lies on a Regge trajectory[7] $\alpha(t)$, given by

$$\alpha(t) = 1 - C_2(A)\frac{g^2}{16\pi^2}(\mu^2 - t)\int_0^1 \frac{d\beta}{\mu^2 - \beta(1-\beta)t} , \qquad (6)$$

$$\sim 1 - C_2(A)\frac{g^2}{8\pi^2}\ln(-t/\mu^2) , \qquad (7)$$

where (7) gives the form for $-t \gg \mu^2$. This is confirmed by a study of the $\mu\,\partial/\partial\mu$ equation (as Carruthers and Zachariasen have independently discovered and reported to this conference). As it is presently formulated (with μ a gluon mass), the $\mu\,\partial/\partial\mu$ equation can only be used in the near-forward regime $s, M^2 \gg |t|$, μ^2 when $|t| \gg \mu^2$, and not when $|t| \simeq \mu^2$, graphs involving the Higgs scalars which generate μ must be saved, but these are not included in the rules given above for forming the $\mu\,\partial/\partial\mu$ equation. However, it may be that if μ is an off-shell infrared cutoff the $\mu\,\partial/\partial\mu$ equation is valid even for $|t| \simeq \mu^2$, and so we indicate the procedure to be used in this case. Consider quark-quark elastic scattering with amplitude $T(s, \vec{\Delta})$, where the momentum transfer t can be expressed in terms of the two-dimensional vertex $\vec{\Delta}$ as $t = -\Delta^2$. Since $\vec{\Delta}$ is not necessarily large compared to the gluon-line integrations the application of the rules for forming the $\mu\,\partial/\partial\mu$ equation yield a convolution, which can be written in two dimensions. There are separate equations for each color representation in the t channel; for the t channel octet amplitude $T^{(8)}$ one finds, when $s, M^2 \gg |t|$,

$$\mu\frac{\partial T^{(8)}}{\partial\mu} = -C_2(A)\frac{g^2}{4\pi^2}H(s/M^2)\int\frac{d^2k}{2\pi}T^{(8)}(s,\vec{k}-\vec{\Delta})\mu\frac{\partial}{\partial\mu}\left(\frac{1}{k^2+\mu^2}\right) . \qquad (8a)$$

In case $|t| >> \mu^2$, k can be dropped compared to Δ in (8a), which then yields

$$\mu \frac{\partial T^{(8)}}{\partial \mu} \to C_2(A) \frac{g^2}{4\pi^2} H\left(\frac{s}{M^2}\right) T^{(8)}, \tag{8b}$$

which is a _local_ equation in momentum space. (8a) is simply solved in the coordinate space conjugate to k (impact parameter space). The limiting form of the solution for $|t| >> \mu^2$ is

$$T^{(8)} = T_B^{(8)} \exp\left\{-H\left(\frac{s}{M^2}\right) C_2(A) \frac{g^2}{8\pi^2} \ln\left(\frac{-t}{\mu^2}\right)\right\}. \tag{9}$$

Of course, (9) also satisfies (8b). For large s, the Born term T_B behaves like s/t, $H(s/M^2) \sim \ln(s/M^2)$ and we recover the infrared-singular gluon Regge trajectory in (7).

It is worth noting that there is a simple interpolation formula which holds both in the infrared limit $(-t \sim s)$ and in the near-forward limit. It is

$$T^{(8)} \simeq T_B^{(8)} G_R^2(t) \exp\left\{-H\left(\frac{s}{M^2}\right) C_2(A) \frac{g^2}{8\pi^2} \ln(-t/\mu^2)\right\}. \tag{10}$$

Recall that G_R is the form factor for R coupled to the gluon (see comments after Eq. (4)), and corresponds to t channel form-factor corrections. With $G_R(t)$ normalized to one at t=0, (10) reduces to (9) for small t.

The differential equation can also be applied to the t channel color singlet amplitude, and may contain information on hadronic physics, e.g., the Pomeron trajectory. At this writing, a detailed comparison with McCoy and Wu's eighth-order results has not yet been made, but

it is simple to do. (There is agreement through sixth
order.) The differential equation predicts Regge cuts,
one moving and infrared-singular, the other fixed at
J=1 and non-singular, as Carruthers and Zachariasen
have shown.

3.   THE KLN THEOREM VERSUS CONFINEMENT: IS PERTURBATION THEORY RELEVANT?

The Kinoshita-Lee-Nausenberg (KLN) theorem states,
in a general way, that suitably defined total (or inclusive) cross-sections do not have mass singularities
in theories with massless particles, when calculated
order by order in perturbation theory. Until recently
this theorem had not been studied for non-Abelian gauge
theories. At the moment, several groups have attacked
the Bloch-Nordsieck problem in Yang-Mills theories, as
reported by Appelquist[8] at this conference. These
authors all agree that the KLN theorem does hold order
by order even for colored processes, and that the cross-
section for emission of undetected soft gluons with a
finite energy resolution is finite. We are in complete
agreement with this result. The question then is: Can
the mass singularities discussed in the first section hav
anything to do with confinement, or do they all cancel in
physical cross-sections?

It appears to us that the $\mu\, \partial/\partial\mu$ equation allows for
the inclusion of <u>non-perturbative</u> effects of mass singularities; these effects are a signal for confinement.
If this is so, then the $\mu\, \partial/\partial\mu$ equation is not just a
way of compactly expressing perturbation theory, but it
can really go beyond an order-by-order expansion. The
effects in question occur even at the simplest level of

application (leading logarithms). We return to Eq. (5),
which essentially expresses the behavior of the amplitude
$T_{(N)}$ for the emission of N gluons with three-momenta
$|\underset{\sim}{k}|>>\mu$. The gluons are on shell $(k^2=\mu^2)$, but the individual components of $\underset{\sim}{k}$ are $>>\mu$. This does not contradict the physical requirement that the gluons be soft,
since we ultimately take μ to be vanishingly small. For
the emission of one gluon of momentum k from $T_{(0)}$, (5)
is integrated as:

$$T_{(1)}^{\alpha}=\hat{T}_{(1)}^{\alpha}\ \exp\{-C_2(A)\frac{g^2}{16\pi^2}\ell n^2\left(\frac{p.k}{M\mu}\right)\}\qquad . \qquad (11)$$

$\hat{T}_{(1)}$ is constructed from $T_{(0)}$ by adding bare gluon
vertices to the external legs; schematically

$$\hat{T}_{(1)}^{\alpha}\sim g\Sigma\frac{p^{\alpha}}{p.k}T_{(0)}\qquad . \qquad (12)$$

The total cross-section for the emission of one gluon
looks like (omitting inessential factors)

$$\sigma_{(1)}\sim g^2\sigma_{(0)}\int\frac{d^3}{2\omega_k(p.k)^2}\ \exp\{-C_2(A)\frac{g^2}{8\pi^2}\ell n^2\left(\frac{p.k}{M\mu}\right)\}\ , \qquad (13)$$

where $\omega_k=(\underset{\sim}{k}^2+\mu^2)^{\frac{1}{2}}$. As it stands, the integral in (13)
can be done without any upper cutoff in k, since the
integrand exponentially drops for p.k>>Mμ. However, if
the integrand is expanded in powers of g^2, each term requires a cutoff, identified (as in QED, where the argument
of the exponent is zero) with the energy resolution ΔE.
With the cutoff supplied, it can be verified that the
total cross-section calculated, say, to $O(g^4)$ is independent of μ, with the real and virtual gluon singularities

exactly cancelling; this is the KLN theorem.

The integral in (13) is finite as $\mu \to 0$, and has a value $\sim g^{-1}$. Thus it is impossible to recover the perturbation expansion (cutoff with resolution ΔE), and we seem to have encountered a genuinely non-perturbative effect. It suggests that the Bloch-Nordsieck program, using a cutoff ΔE, is really unnecessary, because the full theory to all orders will have a finite limit as $\Delta E/\mu \to \infty$. The cross-section for emission of N gluons will be proportional to $\sigma_{(0)}$, with an unknown finite coefficient function. Since for colored processes $\sigma_{(0)}$ vanishes as $\mu \to 0$, we could conclude that the cross-section for any colored process vanishes in this limit. We do not take seriously the precise form of the damping exponential in (13), but (as discussed in the next Section) we speculate that there will be some form of damping of emission of gluons which defeats the application of the KLN theorem to non-Abelian gauge theories.

There are other theories which presumably suffer an analogous failure to the KLN theorem. For example, in 2+1-dimensional QED, the static photon-exchange potential behaves like $\ln r$ for large r which suggests that the charged fermions are confined. The KLN theorem nonetheless predicts the complete cancellation of all photon mass singularities, order by order. However, confinement in 2+1-dimensional QED will be expressed as anomalous behavior of the fermion propagator, which can be fully appreciated only non-perturbatively.

4.   IMPLICATIONS FOR HADRONIC PHYSICS, AND PROGNOSTICATION

What is the relevance of these considerations to hadronic processes? We emphasize three main points:

(A) In the idealization of mimicking bound states as
<u>free</u> clusters of quarks and gluons, localized to a small
space-time region (so that the constituents of the cluster
have essentially the same velocity), the formulas given
above hold, except that the Casimir eigenvalues refer to
the cluster taken as a whole, as if it were an elementary
particle. Of course, for hadrons all the Casimir
eigenvalues are zero, with the consequence that there
are no leading infrared singularities for any hadronic
process. Then (aside from weak ultraviolet singularities
of the type usually found in asymptotically free theories)
the quark-counting rule[9] holds at large momentum transfer
for hadronic form factors and fixed-angle exclusive
processes, since the quark-counting rule reflects the
behavior of the connected Born term. It is worth noting
in this respect that the infrared effects suppress the
Landshoff-graph contributions to wide-angle exclusive
hadronic cross-sections which would violate the quark
counting rule. (B) As discussed in Section 3, the
application of the $\mu \ \partial/\partial\mu$ equation to the emission of
soft gluons yields a <u>signal</u> for confinement: the S-matrix
for a process involving even one on-shell colored particle
vanishes as $\mu \rightarrow 0$. Such a result is necessarily non-
perturbative, and differs greatly from the Bloch-Nordsieck
results in QED. (C) There is some hope that the $\mu \ \partial/\partial\mu$
equation may contain information about the Pomeron. This
comes from the study of the t channel color singlet
amplitude in the near-forward regime. It appears to have
an infrared-singular, moving Regge cut due to the exchange
of two gluons in a non-hadronic state, plus a non-singular,
<u>fixed</u> (at J=1) Regge cut which may be relevant to the
Pomeron. Details are given by Carruthers and Zachariasen
in these proceedings.

Of course, these implications for hadronic physics
are based only on leading-logarithm results, which
hardly give a convincing picture of confinement. The
task for the future is to extend the $\mu \, \partial/\partial\mu$ equations
far beyond the realm of leading logarithms. In so doing,
it becomes necessary to dispose of the crude device of
using μ as a small mass, since there is not supposed to
be any mass for the gluons. The interpretation of μ
as a measure of distance from the mass shell is far
better. It can be done nicely in space-time, by looking
at Green's functions for particles separated by large
distances. (This is closely akin to the momentum-space
techniques already alluded to.[5]) We give an example
from QED to illustrate the technique. A source $J(z)$,
localized in space-time, emits a charged fermion and its
antiparticle. This process is described by the Green's
function

$$G(x,y,z)=<0|T(J(z)\psi(x)\bar{\psi}(y))|0> \, , \qquad (14)$$

which differs from the Fourier transform of the vertex
function for J by having external-leg propagators. Take
$x^o \sim y^o$, and $(\underset{\sim}{x}-\underset{\sim}{z})^2=r^2=(\underset{\sim}{y}-\underset{\sim}{z})^2 \to \infty$. In a theory with only
short-range forces, G would vanish like r^{-1}, just as the
asymptotic Schrödinger wave function would in non-relativistic theory. (This behavior is modified in the presence of long-range forces; if there is confinement, G
will vanish faster than r^{-1}.)

Now let $J(z)$ correspond to a reasonably well-defined
(within the constraints imposed by space-time localization) total momentum P. As the particles recede from
J, it is possible to define their momenta $p,p'(p+p'=P)$
with increasing accuracy, and in the limit of infinite

separation the usual on-shell vertex (if it exists) can
be extracted from G. At the same time their positions
may also be specified, within the uncertainty principle
limits (which are unimportant for separations much
greater than a Compton wavelength). Under these circumstances the infrared-singular part of G is extracted
by using conventional eikonal techniques, the details
of which we omit. The result is that (localizing $J(z)$
at $z\tilde{\,}0$ for convenience)

$$G(x,y,z) \sim S(x)S(y) \; \exp\{ig^2 \int_a^{x \cdot P} ds \int_a^{y \cdot P'} ds' v' \cdot v D(vs-v's')\} \;,$$

(15)

where $S(x)$ is the free fermion propagator, $v_\alpha = p_\alpha M^{-1}$
is the four-velocity of the fermion at x, and similarly
for $v'_\alpha (v^2 = v'^2 = 1)$. $D(z)$ is the photon propagator $\sim z^{-2}$.
The lower limits a on the proper-time integrals in (15)
act as ultraviolet cutoffs and reflect the imperfect
localization of the source. The integrals are readily
carried out; the exponential in (15) becomes

$$\Gamma = \exp\{-\frac{g^2}{4\pi^2}H\left(\frac{t}{M^2}\right)\ell n(x \cdot p + y \cdot p')\} \;,(t=(p+p')^2) \;,$$

(16)

where uninteresting terms are omitted. A factor of i
has disappeared in going from (15) to (16), because the
singular part of the integral responds to the imaginary
part of the photon propagator. (The usual Coulomb phase
comes from the imaginary part of H, which is singular at
the threshold $t=4M^2$.)

It is evident from (15) that the mass-shell limit
of the vertex function Γ corresponds to $x \cdot p \to \infty, y \cdot p' \to \infty$;
if this limit could be taken the exponent of (15) would
become the usual one-loop graph. This limit does not

exist, but we can write a differential equation for Γ
for large but finite x.p, y.p':

$$\left(x\cdot\frac{\partial}{\partial x}+y\cdot\frac{\partial}{\partial y}\right)\Gamma = \frac{g^2}{4\pi^2}H\left(\frac{t}{M^2}\right)\Gamma \tag{17}$$

This is, of course, the space-time version of the
$\mu \, \partial/\partial\mu$ equation (3a) (with $C_2=1$, as appropriate for
QED). The other equation in Section 1 can be similarly
generalized.

In QED, Γ as defined by (16) or (17) vanishes like
a power of $r=|x|$ or $|y|$ for large r; in a theory with
short-range forces it would be independent of r in this
limit. This does not, of course, mean that there is
confinement in QED. In fact, replacing the operator
$\psi(x)\overline{\psi}(y)$ in (14) by the gauge-invariant operator.

$$\psi(x)\overline{\psi}(y) \; \exp\{ig\int_x^y dz^\mu A_\mu\} \quad ,$$

which amounts to adding photons in coherent states,
exactly cancels the infrared effects in (16). Such a
procedure cannot be used in QCD, if the gluons are
confined, so that the coherent states cannot cancel the
virtual-gluon singularities.

The exponential form (15) is reminiscent of a
semi-classical (first-quantized) picture, in which the
asymptotic wave function is approximated by exponentiating
the classical action. The photon propagator D yields the
potential. Confinement folklore holds that the potential
between quarks should grow like r, corresponding to
$D\sim q^{-4}$ in momentum space. So the next step in the program
we outline here is to study the gluon propagator (which
amounts to saving higher-order corrections), to find how

it is modified. As far as we know now, even though the
fermion propagators and fermion vertices may also have
severe infrared singularities (near mass-shell), these
mutually cancel in constructing Green's functions, except
for certain singularities which can be associated with
the gluon propagator. This is not the same as saying
that $Z_1 = Z_2$, which is not true in a general gauge. However,
note that there are gauges (the ghost-free gauges
$n^\alpha A_\alpha(x) = 0$ for some fixed vector n^α) in which $Z_1 \equiv Z_2$, and
these are particularly useful in discussing infrared
problems.[10]

We should study not the conventional gluon propagator
but rather a gauge-invariant propagator

$$i \ D_{\mu\nu} = <0|A_\mu(x) \ P \ \exp\{ig\int_x^y dz^\beta A_\beta\}A_\nu(y)|0> \qquad (18)$$

where the P stands for path-ordering. $D_{\mu\nu}$ depends on the
path chosen; let us take a straight line from x to y.
Then $D_{\mu\nu}$ can be calculated without the exponential factor,
by going to a ghost-free gauge $n^\alpha | (x-y)^\alpha$. In such a
gauge the exact relation

$$\beta(g) = g\gamma_V(g) \qquad (19)$$

holds (just as in QED) in consequence of $Z_1 = Z_3$ (for
gluons); here γ_V is the gluon anomalous dimension. The
relation (19) considerably simplifies the analysis of $D_{\mu\nu}^{-1}$
with renormalization-group techniques; it can be shown that
the scalar self-energies Π (e.g., the coefficient of $g_{\mu\nu}$
in $D_{\mu\nu}^{-1}$) are functions only of the running coupling constant
$\bar{g}$, when fermion loops are omitted. (Massive fermion loops
do not contain important infrared singularities.) More
remarkable, the leading term in the expansion of Π looks

like

$$\Pi(q^2) = q^2 \frac{g^2}{\bar{g}^2(q^2)} \qquad (20)$$

In other words, the perturbation expansion of Π may be a power series in $\bar{g}^{-2}$, _not_ $\bar{g}^{+2}$. This fact may prove very useful for strong-coupling approximations.[11]

In a confined theory, we expect that $\ln \Pi(q^2)=0$ for sufficiently small q^2, since real gluons do not exist. This suggests that $\bar{g}^{-2}(q^2)$ is a polynomial in q^2 for small q^2. Aside from the free-field case $\bar{g}\sim$ constant, the next simplest choice is $\bar{g}^{-2} \sim q^2$, which yields the confining potential $V(r)\sim r$.

The idea, then, is that the general structure of the differential equation (1) will be maintained, except that the operator $\mu \; \partial/\partial\mu$ will be given a new interpretation in space-time. The anomalous "dimensions" Γ are probably given by one-loop skeleton graphs, just as in the perturbative leading-logarithm calculations, except that the gluon effective propagator is considerably modified from the free-field value. The hard job ahead is to calculate these propagator modifications.

ACKNOWLEDGMENTS

One of us (JMC) would like to acknowledge helpful conversations with P. Carruthers and F. Zachariasen on the application of the $\mu \; \partial/\partial\mu$ equation to near-forward scattering.

REFERENCES

1. J. M. Cornwall and G. Tiktopoulos, Phys. Rev. Letters $\underline{35}$, 338 (1975).

2. J. M. Cornwall and G. Tiktopoulos, UCLA preprint TEP/75/21, October 1975 (submitted to Physical Review).

3. J. J. Carrazone, E. R. Poggio, and H. R. Quinn, Phys. Rev. $\underline{D11}$, 2286 (1975). This work on off-shell form factors contains some errors which are corrected in a later preprint.

4. E. R. Poggio and H. R. Quinn, Harvard preprint (1975). This work is on fixed-angle amplitudes for slightly off-shell particles.

5. To construct the nearly-on-shell amplitude, first take all the external momenta on shell: $p_i^2 = M_i^2$ (=0 for a gluon). Then change the momenta to $p_i + k_i$ ($\Sigma k_i = 0$), where the components of the k_i are $O(\mu)$. The differential equation is derived by applying the operator $\Sigma k_i \cdot \partial / \partial k_i$, considering the possible routings of the k_i through the lines of the graph. Details will be given in a forthcoming paper.

6. A pioneering work is by H. T. Nieh and Y. -P. Yao, Phys. Rev. Letters $\underline{32}$, 1074 (1974) (plus a later preprint), but other workers in the field do not agree with these authors. There is agreement among Lipatov; McCoy and Wu; Tyburski; and Cheng and Lo that the gluon Reggeizes, according to recent preprints.

7. As predicted earlier by M. T. Grisaru, H. J. Schnitzer, and H. S. Tsao, Phys. Rev. $\underline{D8}$, 4498 (1973). There is a close connection between Reggeization of an elementary particle and its coupling to gauge

fields, based on the gauge freedom in choosing the wave-function renormalization constant Z_2; see J. M. Cornwall, Phys. Rev. <u>182</u>, 1610 (1969).

8. In addition to a preprint by Appelquist, Carrazone, Roth, and Kluberg-Stern, we have seen preprints by Yao and by Sterman. Other references are given by Appelquist in these proceedings.

9. S. J. Brodsky and G. Farrar, Phys. Rev. Letters <u>31</u>, 1153 (1973); V. A. Matveev, R. M. Muradyan, and A. N. Tavkhelidze, Lett. Nuovo Cimento <u>7</u>, 719 (1973).

10. One of us (JMC) has remarked that there are new singularities in the Feynman integrals of the light-cone gauge ($n^2=0$); see Phys. Rev. <u>D10</u>, 500 (1974). It now appears that these singularities cancel out in Green's functions.

11. Some very interesting results have also been gotten by Migdal, in the context of Wilson's lattice-gauge theory, and by Brezin and Zinn-Justin for the non-linear σ model in 2+ϵ dimensions, which Migdal's work suggests has a similar infrared structure to a gauge theory in 2(2+ϵ) dimensions.

FREEDOM NOW: A NEW LOOK AT THE IMPULSE APPROXIMATION

H. David Politzer*

Lyman Laboratory

Harvard University

Cambridge, Massachusetts 02138

ABSTRACT

Recent work on a field-theoretic impulse approximation is described. The operator product expansion is organized according to powers of $g(Q^2)$, the effective coupling constant appropriate to momentum transfer Q^2, instead of asymptotic dimensional analysis. This allows the calculation of all quark and target mass dependence of lepton-hadron inclusive scattering, order by order in $g(Q^2)$. To $O(g^0)$, there is scaling in a mass dependent variable, ξ. The correct incorporation of quark masses into the renormalization group analysis is presented; the consequences include the relation of constituent masses to current algebra and the magnitude and shapes of the parton distribution functions for gluons, anti-quarks, and heavy quarks.

---

*Junior Fellow, Harvard University Society of Fellows

A simple example of the impulse approximation is provided by the scattering of low energy electrons off hydrogen gas. At what momentum transfers, Q^2, can the process be approximated by scattering off free particles? It is certainly sufficient for Q to be much greater than the proton mass or even the electron mass. In fact, Q need only be much greater than 13.6 eV. That is because there is a small coupling constant, and the struck particle must be approximately free on the length scale set by Q. The impulse approximation has really nothing to do with dimensional analysis or small masses.

In the case of inclusive lepton-hadron scattering (fig. 1) as described by colored-quark-gluon gauge theories, the effective coupling g(Q) must be small (fig. 2) for the applicability of the impulse approximation at momentum transfer Q. Since Q need not be large compared to various masses in the problem, the full mass (quark and target) dependence of lepton-hadron inclusive scattering may be computed by an expansion in g(Q).[1]

The only way I know of implementing these ideas in detail (with control over approximations, etc. and generally knowing exactly what I am talking about) is to use the operator product expansion with the renormalization group. (Some of the results have obvious parton picture analogs; some do not.) I want to emphasize that we use the exact same tools and assumptions used to derive approximate scaling as Q goes to infinity from asymptotic freedom.[2] This work stands on as firm or as shakey footing. We have simply pressed on to extract more detailed (and more interesting) consequences.

There are really two main ideas by which we make the older approach more precise. First, we clarify the quark

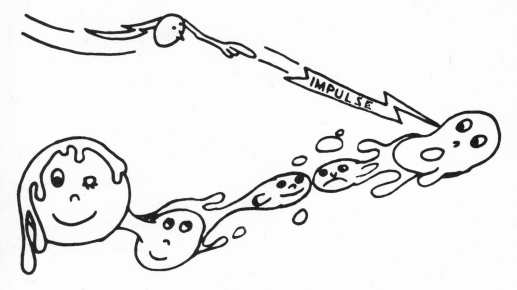

Figure 1

Figure 2

mass dependence of the renormalization group analysis and
introduce definitions appropriate to the problem of
lepton-hadron scattering. (I will say more about this
later.) Second, we organize the operator product expansion according to powers of $g(Q)$, which is small for Q
sufficiently large; from experiment we conclude that
1 GeV is certainly large enough to begin to use this
analysis. This is in contrast to the more traditional
organization according to dimensional analysis, which is
useless if one needs to determine various mass dependences

Given so short a time, I would like to offer just a
sampler of some of our results. I hope there is at least
one thing to interest everyone, so that you will be interested in following the derivation I will outline and
motivated to have a look at our papers.[1] Also, Professor
De Rújula has given an excellent exposition of one of the
most important applications of these ideas earlier at this
conference.

In fig. 3 I show quark masses, m_i, as a function of
scale, M. In the parton language, the effective quark
mass as determined by the response to a momentum transfer,
Q, depends slowly on Q. (In the graph, take Q for M.)
The field theory provides differential equations for $m(M)$.
We must provide boundary conditions. I have taken
$\frac{g^2}{4\pi}(3 \text{ Gev})=1/2$ as a plausible possibility; the charmed quark
mass, $m_{p'}$, is taken to be 1.5 GeV on the scale where it is
measured to be 1.5 GeV, i.e. at M=3 GeV.[3] Similarly, I
think it reasonable to take m_λ (1 GeV)=0.5 GeV. If there
are yet heavier quarks, the mass of the first $q\bar{q}$ vector
meson would give approximately twice the quark mass at the
meson mass. For the light quarks, we go to very large M
where all quark masses go to zero but in the ratios of the
bare masses. Hence, the ratios of $m_i(M)$ for $M \gg m_i$

determine the symmetries of the hadronic Hamiltonian.
PCAC implies that $m_\pi^2/m_k^2 = m_{p,n}/m_\lambda$, which we use as a last
boundary condition. The interesting features of the
resulting curves are: 1) Charm PCAC is a bad approximation; it implies that $m_D \approx 1$ Gev, much too small, 2)
In the domain of interesting experiments, the light quarks
are very light and can be well approximated as massless.
3) The mass of the ρ is crudely predicted to be 500-800
MeV and depends on the coupling constant once given
m_π, m_K and m_ϕ. It is the correct slicing of the curves
that allows us to conclude that $m_\lambda/m_{p,n} \approx 1.5$ (which then
gives m_ϕ/m_ρ), taken as constituent masses (intersecting
$2m=M$) while $m_\lambda/m_{p,n} \approx 20$ in the Hamiltonian.

Fig. 4 shows the results of a calculation of the
integrals of various distribution functions (or the
contribution to the total energy momentum of the target).
As input, I assumed that the antiquark and heavy quark
distributions are very small at low Q(eg 1 Gev). I also
use the experimental result that half the energy-momentum
is carried by glue. If there are yet heavier quarks,
they would increase slower than the p' curve.[4] I plotted
the curves until asymptotic SU(4) sets in just to emphasize how far away and unphysical that limit is.

We can also conclude something about the shapes of
the distributions for $\xi > 1/3$. In parton language if we
assume that there are few hard gluons, hard anti quarks,
and hard heavy quarks, those that there are come off the
hard valence quarks. We find that if the valence distribution vanishes like $(1-\xi)^a$ for $1/3 < \xi \lesssim 1$, then the glue
goes like $(1-\xi)^{a+1}$ and the sea quarks go like $(1-\xi)^{a+2}$
(or one power faster if $m(Q) >> Q$.)

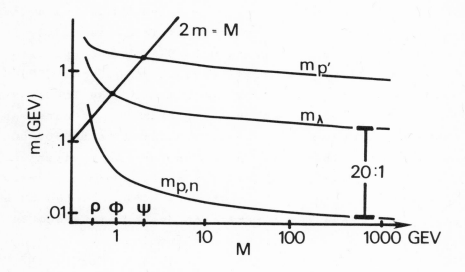

Figure 3

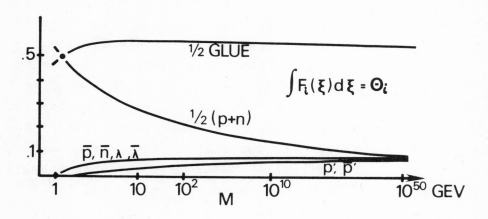

Figure 4

From the reorganization of the operator product expansion to zeroth order in g we get scaling in a mass dependent variable, ξ.

$$\xi = \frac{Q'^2}{2m_p\nu} \quad \frac{2}{1+\sqrt{1+Q'^2/\nu^2}} \quad , \tag{1}$$

$$Q'^2 = \frac{1}{2}\{Q^2+m_F^2-m_I^2 + \sqrt{Q^4+2Q^2(m_F^2+m_I^2) + (m_F^2-m_I^2)^2}\} \quad ,$$

where m_p is the target mass, m_I is the struck quark mass, m_F is the produced quark mass, and Q^2 and ν are as usual.

In a process where both m_F and m_I are negligible, ξ is sort of half way between x and x', the Bloom-Gilman variable. Note, however, that for Q comparable to m_p, the $O(g^2(Q))$ corrections to ξ-scaling are non-negligible and in fact are of the right order of magnitude and structure to make up for the difference between ξ and x'.

An example of striking a heavy quark which remains heavy is given by scattering a muon electromagnetically off a virtual charmed quark. For such a case, as Q increases for fixed x, ξ decreases. If the charm distribution is a decreasing function of ξ, then at fixed x the observed structure function would increase with Q. The observed non-scaling in μ-p scattering is of the opposite direction and hence not explainable by charm and rather is evidence for the $O(g^2)$ effects, which have that signature.

When a light quark is converted to a heavy one via a weak current, $\xi \approx x \frac{Q^2+m_H^2}{Q^2}$. Until Q^2 $(=2m_p E_\nu xy)$ is much larger than m_H^2, x is rather smaller than ξ. If valence quarks are struck, with a mean ξ of about 1/4, the mean x will be much smaller until well above charm threshold.

The expected y distributions are also radically altered near charm threshold as implied by scaling in ξ (not x).

The turning of heavy quarks into light quarks can also be analyzed, and $\xi \approx x$. However, as mentioned above, the probability of finding a heavy quark is small at accessible energies and vanishes very rapidly as $\xi \rightarrow 1$.

The renormalization scheme that we claim is optimal from the standpoint of analyzing quark mass effects (or at least representative of an optimal class of conventions is as follows: Define the quark mass by normalizing the inverse propagator to its free field form at a Euclidean momentum, i.e., $S^{-1}(P^2=-M^2)=\not{p}-m(M)$. The coupling constant is defined by some three point function at particular momenta, all of scale M, but including quark mass effects. The first outstanding virtue of these definition is that they make the theorem of Appelquist and Carazzone[5] manifest: if all external momenta are of order M, then all graphs containing heavy particles, $m_H >> M$, are suppressed by M^2/m_H^2. Secondly, in this framework small g implies the calculability of spacelike Green's functions and _vice_ _versa_. That is to say there are never any large logarithms like $\log m^2/M^2$ for $m^2 >> M^2$ or $\log P_i P_j/m^2$ for $M^2 >> m^2$. Such is not the case in virtually all renormalization schemes previously considered. In greater detail, this method gives a description of the transition from $m >> M$ to $m << M$. Finally, m(Q) is precisely the quark mass in the parton model, which when justified by field theory, has a slow Q dependence; that is to say that in the leading impulse approximation the produced parton must have its four-momentum-squared, $(\xi p+q)^2$ equal to $m^2(\sqrt{-q^2}=Q)$.

I would finally like to describe our proposed organization of the operator product expansion. Schematically, the inclusive differential cross section for lepton-hadron

scattering is determined by the target matrix element
of the product of two currents, which is expanded in terms
of a complete set of operators, O^n with coefficients
C_n:

$$<P|JJ|P> = \sum_n C_n \left(Q, g(Q), m(Q)\right) \exp\left[\int_M^Q \gamma_n \frac{dM'}{M'}\right]$$
(2)

$$\times \quad <P|O^n(M)|P>$$

The utility of this form is that all of the Q dependence
is in the C_n, which are calculable in perturbation theory
if $g(Q)$ is small. The C_n are computed using Q as the
renormalization scale. The O^n are renormalized at M,
chosen to be roughly the target mass so that dimensional
analysis order of magnitude estimates of $<P|O^n(M)|P>$ can
be made. The exponential factor is required by the renormalization group to connect renormalizing C_n at Q and
O^n at M. To make practical use of this analysis, the
operators must be ranked according to importance. I
divide this into three steps. 1) Throw out all O^n with
$\not{D}$ or D^2 (made from the gauge covariant derivative) acting on a quark field. This in no way compromises the
completeness of the list of operators because the existence of operator field equations guarantees that these
operators are not linearly independent of the operators
remaining in the list. The remaining operators have the
property that none of them have anomalously large matrix
elements for $m >> m_p \sim M$ (in contrast to the thrown out
operators). 2) Make all the O^n traceless. This intro-
duces no loss of generality as the traces reappear as
$g_{\mu\nu}$'s in the coefficient functions times operators of
lower spin. This ensures that the target mass dependence
(as it enters the Q dependence) of the operator matrix
elements is determined completely by the tracelessness

condition.   3) Finally, order the operators by the magnitude of their coefficient functions in powers of $g(Q)$.
We do <u>not</u> use perturbation theory to compute matrix
elements; rather, all things being equal, we assume all
matrix elements are of the same order of magnitude.   But
in the expansion of the product of two particular currents
these operators enter with coefficients determined by
$g(Q)$.

To order g^o, only quark bilinears have non-vanishing
coefficients, and we are led to investigate the free-
field operator product expansion, keeping all mass dependence.   This approximation leads to ξ-scaling.

I would like to say some final words concerning the
derivation of ξ- scaling because, while conceptually
straight forward, it is technically very complex.   To
order g^o, we study the matrix elements.

$$<P|\bar{\psi}(\partial_{\mu_1}\ldots\partial_{\mu_n} - \text{traces})\psi|P> =$$

$$= A_n\left(P_{\mu_1}\ldots P_{\mu_n} - \text{traces}(e.g. 1/4g_{\mu\nu}m_p^2)\right). \qquad (3)$$

The A_n can be organized into a function G whose moments
are A_n:

$$A_n = \int_o^1 G(Z)Z^n \, dZ. \qquad (4)$$

However, G is <u>not</u> the measured target structure function.
For example, the nth moment of the structure function
depends on all A_m for $m \geqslant n$.   However from a measurement
of the structure function at a single Q one can deduce
G.   Then the subsequent Q evolution is determined by the
$C_n(Q)$.   I should also note that the computation of the
C_n for the case of a heavy struck quark is non-trivial
in free field theory after performing steps 1) and 2).

To summarize, we have given a refined analysis of

lepton-hadron scattering which uses a systematic expansion in the effective coupling $g(Q)$, but to each order it
keeps all mass dependence. Really no new assumptions are
introduced beyond those used to derive approximate asymptotic scaling. This analysis is essential to any interpretation of low Q or heavy quark threshold problems in
lepton-hadron physics.

REFERENCES

1. H. Georgi and H. D. Politzer, "Precocious Scaling,
    Re-Scaling, and ξ-Scaling," and "Freedom at Moderate
    Energies: Masses in Color Dynamics" (Harvard pre-
    prints).

2. H. D. Politzer, Physics Reports $\underline{14C}$, 130 (1974) and
    references therein.

3. T. Appelquist and H. D. Politzer, Phys. Rev. Lett.
    $\underline{34}$, 43, (1975) and Phys. Rev. $\underline{D12}$, 1404, (1975).

4. E. Witten has an alternative but essentially equi-
    valent derivation of these results (to be published)
    based on an expansion in $1/m_H^2$.

5. T. Appelquist and J. Carazzone, Phys. Rev. $\underline{D11}$ 2856,
    (1975).

RELATIVISTICALLY INVARIANT LATTICE THEORIES*

Kenneth G. Wilson

Laboratory of Nuclear Studies

Cornell University, Ithaca, New York   14853

It is shown that quantum field theories on a
Euclidean space time lattice can be derived which
have the same S matrix (after analytic continuation to
the Lorentz metric) as a relativistic continuum theory.
The lattice spacing of the lattice theory is fixed; no
limit of zero lattice spacing is required. It is proposed that (as already discussed by Migdal) a lattice
theory be derived as an intermediate step in solving
the asymptotically free color gauge theory.

It is evident that the Feynman graph expansion will
not be sufficient to demonstrate quark confinement and
compute the bound state spectrum of the colored quark
gauge theory. An alternative framework exists in which
quark confinement is explicit and the only physical
states are color singlets. This alternative framework
is the lattice gauge theory in the limit of very strong
coupling. The strong coupling approximation to the
lattice color gauge theory is reviewed in Refs. 1 and 2.

---

*Supported in part by the National Science Foundation

The strong coupling theory described in Refs. 1 and 2 has one vital flaw: it requires a lattice spacing of order 1/1 GeV. For smaller lattice spacings the particle masses are too heavy to fit experiment. The large lattice spacing means the theory is not Lorentz invariant. It is not even rotationally invariant. In addition there are mass differences (π-η, and N-Δ) which are much too small to fit experiment.

There is, in principle, a resolution to this problem. Namely, a lattice theory can be constructed with a fixed lattice spacing (say 1/1 GeV) such that it has <u>exactly the same S matrix as the continuum asymptotically free gauge theory</u>. That is, despite the large lattice spacing, the physical particle masses and particle scattering amplitudes are the same as for the continuum theory. This will be demonstrated in this talk. However, to keep the exposition simple, only the theory of a single scalar field will be discussed. The complications caused by spin and gauge invariance will not be discussed, except at the end of this talk.

A price has to be paid to achieve the simplicity of a lattice with fixed spacing. The price is that the Lagrangian of the lattice theory is an effective Lagrangian containing terms of arbitrary complexity: 4 quark couplings, 6 quark couplings, 20 quark - 16 gluon couplings, etc. This is in contrast to the simple bare Lagrangian used in the previous strong coupling calculations[1,2]. However, it is hoped that the dominant couplings in the effective Lagrangian involve only 2 to 4 particles. This will be discussed later. This is only a hope: the author has not completed the calculations needed to test this hope.

To introduce the lattice one must break the construction of the solution of the asymptotically free gauge theory into two parts. The first part is the introduction of the lattice and the construction of the effective Lagrangian on the lattice. The second part is the solution of the lattice theory. Before introducing the lattice explicitly the basic idea of the two-step calculation will be explained.

Assume, as in the gauge theory, that one has a theory with a zero mass propagator $1/p^2$. Write this propagator in two parts, a high momentum part and a low momentum part (see fig. 1). For example, (we use a Euclidean metric here)

$$\frac{1}{p^2} = \frac{1}{p^2+\Lambda^2} + \frac{\Lambda^2}{p^2(p^2+\Lambda^2)} \quad , \qquad (1)$$

where the high momentum part (labelled H in fig. 1) is $\frac{1}{p^2+\Lambda^2}$: for $p \gg \Lambda$ this is almost equal to $1/p^2$, while for $p \ll \Lambda$ the high momentum part lacks the zero mass pole for $p^2 = 0$. Conversely the low momentum part $\Lambda^2/[p^2(p^2+\Lambda^2)]$ is almost $1/p^2$ for $p \ll \Lambda$ but is strongly cutoff for $p \gg \Lambda$. The parameter Λ can be chosen at will; for strong interactions one might choose $\Lambda = 1$ GeV (see later).

Consider now the diagram shown in Fig. 2a. The two-step calculation of this diagram consists in breaking it down into subdiagrams containing all possible combinations of high (H) and low (L) momentum parts of the propagator. An example of a subdiagram with four high momentum parts and one low momentum part is shown in Fig. 2b. Then the first step is to compute the part

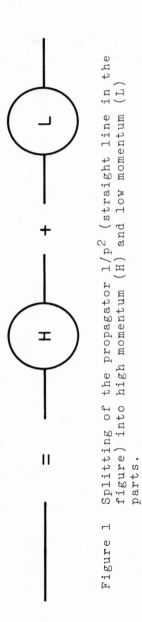

Figure 1   Splitting of the propagator $1/p^2$ (straight line in the figure) into high momentum (H) and low momentum (L) parts.

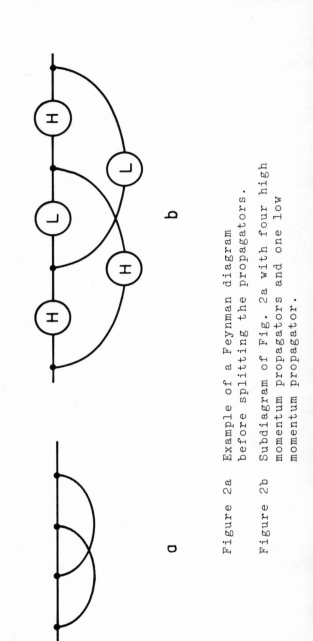

Figure 2a   Example of a Feynman diagram before splitting the propagators.

Figure 2b   Subdiagram of Fig. 2a with four high momentum propagators and one low momentum propagator.

(or parts) of each diagram containing the high momentum propagators. The result of the first step is a set of effective vertices. For example, the diagram of Fig. 2b leads to an effective vertex with four external lines (see Fig. 3). The second step of the two step calculation is to combine the effective vertices and low momentum propagators to complete the diagram. In the example of Fig. 2b, this means combining two legs of the effective vertex of Fig. 3 with a low momentum propagator (see Fig. 4).

In summary the second step of the calculation is a calculation based on a set of Feynman rules but using the effective vertices instead of the bare vertices and the low momentum part of the propagator instead of the full propagator. The first step of the calculation is the calculation of the effective vertices from the bare vertices. The first step is again a Feynman graph calculation; it uses bare vertices and the high momentum part of the propagator. The idea of using a two step calculation was proposed by Dyson long ago[3].

In the case of the asymptotically free gauge theory, the purpose of the two step calculation is to separate the weak coupling calculation and ultraviolet renormalization from the strong coupling calculation and infrared divergence problems. The first step of the calculation using high momentum parts of the quark and gluon propagators has no infrared divergences because the high momentum propagators have infrared cutoffs. As long as Λ is large enough (hopefully of order 1 GeV or so) the first step of the calculation can be done by perturbation theory without having to sum large numbers of graphs. This is due to the small

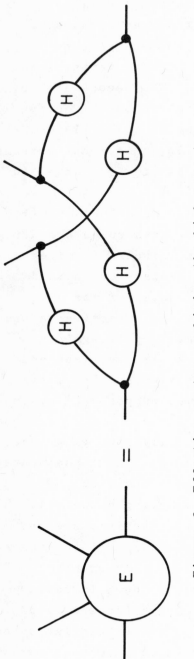

Figure 3   Effective vertex combining the high momentum
            propagators of Fig. 2b.

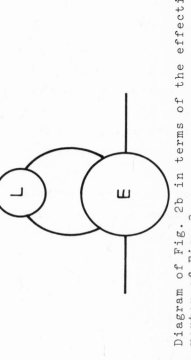

Figure 4   Diagram of Fig. 2b in terms of the effective
            vertex of Fig. 3.

effective coupling for p > 1 GeV. Furthermore, the
effective vertices will decrease in size as the number
of external legs increases, because effective vertices
with a large number of external legs involve a large
power of the effective coupling constant. The standard
ultraviolet divergences are present in the first stage
of the calculation; these are removed by standard
renormalization methods. The effective vertices are
(after renormalization) free of divergences. No
divergences arise in the second step of the calculation
because the low momentum propagator is cut off at
large p (if the $1/p^4$ dependence at large p in Eq. 1
still causes singularities one can make a high-low
momentum split such that the low momentum propagator
behaves as $1/p^6$ or $1/p^8$ or faster for large p).

The infrared divergence problems return in the
second stage of the calculation. The propagators of the
second stage have zero mass poles, causing infrared
divergences. The difficulties of these divergences are
not diminished by the two step calculation. Hence the
lattice will be introduced which (hopefully) allows
the use of strong coupling lattice methods. Within the
strong coupling expansion there are no infrared
divergences: see refs. 1 and 2.

Now the lattice will be introduced. The lattice is
a hypercubic lattice (in a Euclidean metric) with spacing
a, where a is of order 1/1 GeV and held fixed. In order
to explain the ideas involved, we shall use perturbation
theory to discuss both steps in the calculation.

There would appear to be obvious contradictions in
a theory on a lattice with a relativistic S. matrix. The
first problem is with the single particle energy-momentum
relation. Relativistically, the relation must be

$$E = \sqrt{\vec{p}^2 + m^2} \, , \qquad (2)$$

where E is the energy, $\vec{p}$ the momentum, and m the particle mass. But on a lattice a plane wave has the form $e^{i\vec{p}\cdot\vec{n}a}$ where $\vec{n}$ is a vector with integral components marking the position of a lattice site. The plane wave is unchanged by a translation

$$p_i \to p_i + \frac{2\pi}{a} \, , \qquad (3)$$

where p_i is any component of $\vec{p}$. On the lattice, therefore, the energy E must also be a periodic function of p_i. The relativistic formula (2) is not periodic.

The resolution of this problem is to allow the single particle spectrum to have an infinite number of distinct branches, obtained by translating the relativistic spectrum by multiples of $\frac{2\pi}{a}$. See Fig. 5.

Using multiple branches one obtains a spectrum completely invariant under $p_i \to p_i + \frac{2\pi}{a}$, while at the same time one of the branches is the exact relativistic spectrum $E = \sqrt{\vec{p}^2 + m^2}$. Note that a is <u>fixed</u>; <u>it is not necessary</u> to take the limit $a \to 0$.

The multiple branches are obtained by having a propagator with multiple poles. The simplest form for the (zero mass) propagator would be

$$D_{lat}(p) = \sum_{\ell} \frac{1}{(p+\frac{2\pi\ell}{a})^2} \, , \qquad (4)$$

where ℓ is a four vector with integer components. Unfortunately, this sum diverges. A more general formula is

$$D_{lat}(p) = c + \sum_{\ell} \frac{[\gamma(p + 2\pi\ell/a)]^2}{(p + \frac{2\pi\ell}{a})^2} \, , \qquad (4)$$

where $\gamma(p)$ is an arbitrary bounded function of p
which goes to zero rapidly for $p \to \infty$. The constant
c is also arbitrary; it provides further flexibility
in defining the lattice theory. $D_{lat}(p)$ is invariant
to translations of any component of p by $2\pi/a$; hence
it is a possible lattice propagator. A 4-dimensional
Euclidean lattice is being used: the metric is
$p^2 = p_0^2 + \vec{p}^2$. The poles of the propagator are obtained
by analytic continuation to the Lorentz metric. The
complete set of poles occur for

$$E = ip_0 = \frac{2\pi\ell_0 i}{a} \pm \sqrt{(\vec{p} + \frac{2\pi\ell}{a})^2} . \quad (5)$$

These poles include all of the branches of the spectrum
proposed earlier plus the additional branches for
complex E due to the lattice in imaginary time.

The factor $\gamma(p)$ is arbitrary; it does not change
the locations of the poles of the propagator and hence
does not affect the single particle spectrum. The
multiple branches of the spectrum also are not a problem
since the physical spectrum can be determined: it is
the branch for which E is real and smallest for $\vec{p} = 0$.
It is logical to choose $\gamma(p)$ to become small when
$|p|>>\pi/a$. Then π/a is the cutoff momentum analogous
to Λ in the formula (1). A typical form

$$\text{for } \gamma(p) \text{ is } \left\{\gamma(p)\right\}^2 = \Pi_\mu \left\{ \frac{4 \sin^2(p_\mu a/2)}{p_\mu^2/a^2} \right\}. \text{ See later.}$$

Now consider the multi-point (vertex) functions.
The vertex functions of a lattice theory must also be
periodic functions of the external momenta in the vertex.
This is true both for the effective vertices after step

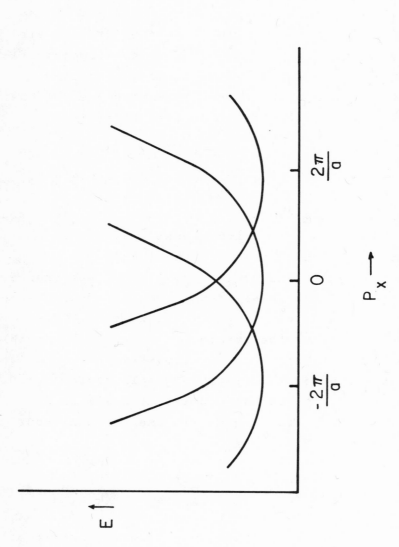

Figure 5   Lattice generalization of the relativistic single particle
spectrum, with an infinite number of branches $E(\vec{p})$.

one (which describe the effective Lagrangian on the
lattice) <u>and</u> for the final vertex functions (vacuum
expectation values) after solving step 2. Hence the
vacuum expectation values of the lattice theory cannot
be the vacuum expectation values of the continuum
theory. However, we can set up a simple conversion
procedure analogous to the procedure used for the
propagator. Namely, for each external leg of a vacuum
expectation value of the continuum theory we form a
discrete sum over ℓ. Let the four-momentum of the
leg be p; the sum is

$$\sum_{\ell} \gamma(p + \frac{2\pi\ell}{a}) \times$$

continuum vacuum expectation value $(p + \frac{2\pi\ell}{a})$.

The factor $\gamma(p)$ is needed for convergence. It is
also present to ensure that the S matrix of the lattice
theory is the same as the $\bar{S}$ matrix of the continuum
theory. The standard (LSZ) rules for extracting S matrix
elements from vacuum expectation values are (1) remove
the single particle pole from each external line, (2)
put the external line on the mass shell, (3) divide by
the square root of the residue of the propagator. Part
(1) is necessary because we are dealing with <u>complete</u>
vacuum expectation values which have propagators on all
external lines. In the case of the lattice theory,
demanding that the external line has a pole picks out a
single term from the sum over ℓ for that external line:
if $p^2 = 0$, then $(p + \frac{2\pi\ell}{a})^2$ is non-zero for all non-zero
ℓ. Hence the selection of the pole term at $p^2 = 0$ removes
the sum over ℓ leaving only the $\ell = 0$ term. Step 3
(of LSZ) removes $\gamma(p)$ since (by definition) this is the
square root of the residue of the propagator for $p^2 = 0$.

Thus the process of obtaining the S matrix gives back
the continuum S matrix.

Now we need the rules for constructing the
effective vertices which are input to step 2. The
effective vertices represent an effective Lagrangian
on the lattice; they must also be periodic in their
external momenta. The external line of an effective
vertex may become an external line of a vacuum expectation value on the lattice; the only change is that
the line is multiplied by the lattice propagator. We
have already specified this propagator. Thus the
effective vertices must include a sum of the form

$$\sum_{\ell} \gamma\left(p + \frac{2\pi\ell}{a}\right) \; \frac{1}{D_{lat}(p)} \; \frac{1}{\left(p + \frac{2\pi\ell}{a}\right)^2} \quad \times$$

$$\times \text{ original effective vertex } \left(p + \frac{2\pi\ell}{a}\right) \; .$$

In diagrammatic terms the effective vertex is
computed by first combining bare vertices and high
momentum propagators, as in Fig. 3. No propagators
are included for external lines. Then the sum over
ℓ given above is performed for each external line. If
the external line becomes an external line of a vacuum
expectation value, the factor $1/D_{lat}(p)$ is cancelled by
later multiplication by $D_{lat}(p)$. The factor
$1/(p + 2\pi\ell/a)^2$ is necessary to complete the external
line of a continuum vacuum expectation value, and the
sum over ℓ times $\gamma(p + 2\pi\ell/a)$ is the sum defined previously that generates the lattice vacuum expectation
value.

The external line sum defined above for effective

vertices can be denoted by an "x" on the external
line (Fig. 6). A consequence of this operation is
that when effective vertices are connected by step 2
(lattice) propagators, the original continuum effective
vertices are connected by a lattice propagator times
two "x" operations (Fig. 6). This means that when
diagrams are broken down into bare vertices, the low
momentum propagator connecting bare vertices is the
lattice propagator between two "x" operations. Thus
the requirement that the sum of the high and low
momentum propagators give the full propagator now
becomes

$$\frac{(2\pi)^4}{p^2}\delta(p - p') = D_H(p, p')$$

$$+\gamma(p)\frac{1}{D_{lat}(p)}\frac{1}{p^2}\sum_\ell(2\pi)^4\delta(p-p'-\frac{2\pi\ell}{a})D_{lat}(p)\frac{1}{p'^2}\frac{1}{D_{lat}(p')}\gamma(p')$$

$$(6)$$

(see Fig. 7). In the last term p and p' can differ
by a multiple of $2\pi/a$, as indicated in the δ-function;
the reason for this is that the two "x" operations
involve independent ℓ sums. In the actual calculations
these are two ℓ sums and the momentum p of the lattice
propagator is restricted to lie in the range $-\pi/a$ to
π/a. But this is equivalent to allowing p and p' to have
an infinite range and then using a single ℓ sum to
determine the possible values of p-p'. The above
equation determines the high momentum propagator:

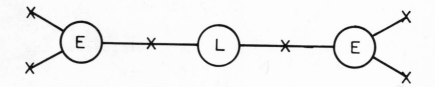

Figure 6   Example of a Step 2 diagram with effective
            vertices converted to lattice vertices by
            the "x" operation defined in the text.

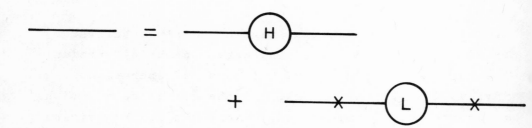

Figure 7   Break-up of the propagator into high and low
            momentum parts in the lattice approach.

$$D_H(p,p') = \frac{(2\pi)^4 \delta^4(p-p')}{p^2} -$$

$$\frac{1}{p^2}\gamma(p)(2\pi)^4 \sum_\ell \delta(p-p'-\frac{2\pi\ell}{a})\frac{1}{D_{lat}(p)}\gamma(p')\frac{1}{p'^2}$$

$$(7)$$

(Note that $D_{lat}(p) = D_{lat}(p')$ because of the periodicity of the lattice propagator).

This completes the specifications of the two steps of defining and then solving the lattice theory. The S matrix and single particle spectrum of the lattice theory include the S matrix and single particle spectrum of the continuum theory. One must select the relativistic branch of the lattice spectrum; this is accomplished when going onto the mass shell by putting $E=\sqrt{\vec{p}^2 + m^2}$ rather than $E=\sqrt{(\vec{p} + \frac{2\pi\ell}{a})^2 + m^2}$. The lattice spacing a is fixed; it can be chosen arbitrarily.

In the case of the asymptotically free gauge theory, the choice of a is governed by the fact that both steps 1 and 2 can only be carried out approximately. It steps 1 and 2 could be done exactly, the S matrix would be independent of a. However, due to the approximations, the actual calculations will be a dependent, and a must be chosen to minimize the cal-culational errors. Because step 1 is proposed to be a perturbative calculation, a must be small enough so that the effective coupling at high momenta (above π/a) is small. Because step 2 is proposed to be a strong coupling calculation, a must be large enough so that the effective coupling at momenta < π/a is large. It is not known yet whether there is any single value for a which satisfies both requirements.

The requirement of gauge invariance complicates the construction of effective vertices for the color gauge theory. These requirements can be met, but the relevant formulae yield complicated Feynman rules. To present these formulae compactly one has to use Feynman path integrals. The formulae for scalar, spinor, and gauge fields will be given here in path integral form. For the scalar case, let $A_{eff}[\phi]$ be the effective lattice action:

$$A_{eff}(\phi) = \frac{a^8}{2} \sum_{n,m} \phi_n \frac{1}{D_{lat}(n-m)} \phi_m + a^{12} \sum_{n,m,k} \phi_n \phi_m \phi_k V(n,m,k) + \dots,$$

$$(8)$$

where $D_{lat}(n-m)$ is the Fourier transform of $D_{lat}(p)$, $V(n,m,k)$ is the Fourier transform of the effective three-line vertex on the lattice (including the "x" operation on all external lines), etc. Then

$$e^{A_{eff}(\phi)} = \int_{\phi'} \exp\left\{ -\frac{a^4}{2} \sum_n c^{-1} [\phi_n - \int_x \gamma(x-na)\phi'(x)]^2 \right\} e^{A_{orig}[\phi]},$$

$$(9)$$

where $\gamma(x)$ is the Fourier transform of $\gamma(p)$ and c is the arbitrary constant added to the lattice propagator (Eq. 4). $\int_{\phi'}$ means a functional integral over all fields $\phi'(x)$. A Euclidean space-time metric is assumed.

A simple choice for $\gamma(x)$ is a step function:

$$\gamma(x) = \frac{1}{a^4} \text{ for } |x_\mu| < a/2 \text{ (all x)} \qquad (10)$$

and 0 elsewhere: this form leads to the specific form of $\gamma(p)$ proposed earlier. With this choice of $\gamma(x)$ the first lattice field ϕ_n is coupled to the average of the

continuum field over a region ("block") of size the
lattice spacing a.

The earlier argument that the S matrix is preserved
was based on perturbation theory. Starting from the
path integral formula a more general argument can be
given. The integrand of the path integral has two
parts: the original action and a "kernel" K:

$$K(\phi,\phi')=\exp\left\{-\frac{a^4}{2}c^{-1}\sum_n [\phi_n - \int_x \gamma(x-na)\phi'(x)]^2\right\} .$$

(11)

The crucial property of the kernel which preserves the
S matrix is the fact that the kernel gives a constant
when integrated over the lattice variables:

$$\Pi_n \int_{-\infty}^{\infty} d\phi_n K(\phi,\phi') = \Pi_n[2\pi c/a^4]^{1/2},$$

(12)

which is a constant. To see that the S matrix is
preserved, one recalls that the S matrix can be derived
from vacuum expectation values of fields well separated
in position space (corresponding physically to detectors
which are microscopically well separated). These
vacuum expectation values can be obtained from a
generating function

$$\Pi_n \int_{-\infty}^{\infty} d\phi_n e^{\sum_n j_n\phi_n} e^{A_{eff}(\phi)},$$

(13)

where j_n is nonzero only in well separated spacetime
regions and one only needs the terms linear in j_n within
a given region. The generating function can be computed
by performing the ϕ integrations inside the ϕ' path
integral: the ϕ integrations become

$$\Pi_n \int_{\infty}^{\infty} d\phi_n e^{j_n\phi_n} K(\phi,\phi') .$$

(14)

If j is 0 this integral is a constant; thus the
nontrivial terms must multiply j. The kernel is a
product $j_n j'_n$, for n' = n. Thus the result has the form

$$\Pi_n \int_{-\infty}^{\infty} d\phi_n \, e^{j_n \phi_n} K(\phi,\phi') = \Pi_n (2\pi c/a^4)^{1/2} \, e^{j_n f[\int_x \gamma(x-na)\phi'(x)]} \tag{15}$$

(to linear order in j_n),

where f is a function of the "block field" $\int_x \gamma(x-na)\phi'(x)$.
The result of the ϕ' path integral turns the ϕ' integral
into the generating functional for vacuum expectation
values in the continuum theory, the interpolating field
being $f[\int_x \gamma(x-na)\phi'(x)]$. Since the S matrix is
independent of the choice of interpolating field, the
lattice theory and the continuum theory have the same
S matrix. This argument is true for a very general
class of kernels: they must be sufficiently local so
that the integral

$$\Pi_n \int_{-\infty}^{\infty} d\phi_n \, e^{\sum_n j_n \phi_n} K(\phi,\phi')$$

does not generate cross-products $j_n j_{n'}$, for large
separations $|n-n'|$, and, most importantly, the integral
must be a constant when $j_n = 0$. (If one obtains a
non-trivial function of ϕ' when j = 0 this function is
added to the original action to make a new action whose
S matrix is equal to the lattice S matrix).

The construction of the lattice action described
here is called the "block spin" method because it was
originally developed in statistical mechanics where the
ϕ variables refer to spins rather than quantum fields[4].
Kanadoff[5] has stressed the requirement that the integral
of the kernel over ϕ be a constant.

It should be noted that vacuum expectation values of currents cannot be determined in the same way the S matrix is obtained. Vacuum expectation values of currents have to be computed by adding a source term $\int j_{\mu ext}(x)j_{\mu}(x)d^{4}x$ to the continuum action inside the block spin path integral, producing a lattice action depending on the external source function $j_{\mu ext}(x)$. In the second stage one computes the lattice path integral over this lattice action. The result is the generating functional for current vacuum expectation values. Block spin kernels very similar to the one of this paper have been studied in detail (in the context of statistical mechanics) by Bell and Wilson[6].

The block spin kernel is easily generalized to the Fermi case: ϕ_{n}^{2} becomes $\overline{\psi}_{n}\psi_{n}$ and $\int_{x}\gamma(x-na)\phi'(x)$ becomes $\int_{x}\gamma(x-na)\overline{\psi}'(x)$ or $\int_{x}\gamma(x-na)\overline{\psi}'(x)$. The path integral over ψ' and $\overline{\psi}'$ involves integration over anticommuting variables. For the gauge field a possible formula is as follows. We use the lattice formalism of Refs. 1 and 2. Define

$$e^{K(W)} = \int_{U}e^{\beta/g_{o}^{2}\text{Tr}(U W^{+} + U^{+} W)}, \qquad (16)$$

where U and W are SU(3) matrices (3 x 3 unitary unimodular matrices) and $\int_{U}$ is the invariant SU(3) group integral. β is an arbitrary constant (like c) and g_{o} is the bare gauge field coupling constant. Define the effective action starting from an initial action on a lattice with very small spacing. Let the original lattice spacing be a/N.

Then

$$e^{A_{eff}(U)} = \int_{U'} \exp \sum_{n\mu} \beta/g_o^2 \left\{ Tr(U_{n\mu}W_{n\mu}^+ + U_{n\mu}^+ W_{n\mu}) - K(W_{n\mu}) \right\}$$

$$\times\; e^{A_{orig}(U')}, \tag{17}$$

where

$$A_{orig}(U') = \frac{1}{g_o^2} \sum_{m\mu\nu} Tr(U'_{m\mu} U'_{m+\hat{\mu},\nu} U'^+_{m+\hat{\nu},\mu} U'^+_{m\nu}), \tag{18}$$

$$W_{n\mu} = \frac{1}{N^4} \sum_{m\epsilon block\; n} U'_{m\mu} U'_{m+\hat{\mu},\mu} U_{m+2\mu,\mu} \cdots U'_{m+(N-1)\mu,\mu}, \tag{19}$$

where "block n" includes the original lattice sites near to the effective site n, for example, the sites m with $-N + 1 \le (m_\mu - n_\mu N) \le 0$ for all coordinate directions μ. The variables $U_{n\mu}$ and $U_{m\mu}$ are SU(3) matrices. The effective action is invariant to gauge transformations where

$$U_{n\mu} \rightarrow V_n U_{n\mu} V_{n+\hat{\mu}}. \tag{20}$$

To show this, one shows that the path integral is invariant to this transformation when the U' variables undergo a "block" gauge transformation, namely

$$U'_{m\mu} \rightarrow V'_m U'_{m\mu} V'^+_{m+\hat{\mu}}, \tag{21}$$

where $V'_m = V_n$ when the old site m is in the block n.

To calculate the effective action in perturbation theory one must add a gauge-fixing term to the original action. To preserve gauge invariance of the effective action, one must modify the gauge fixing term so that it is invariant to block gauge transformations while still fixing other gauge transformations. The resulting expressions are complicated and will not be reported here.

A crude (but calculable) block spin formalism for gauge fields has been developed by Migdal[7]. Migdal's formula is useful for studying the crossover from weak to strong coupling. Improved forms of Migdal's formula are being developed by Kadanoff[8].

ACKNOWLEDGEMENT

I wish to thank K. Subbarao and B. Baaquie for discussions.

REFERENCES

1. K. Wilson, Erice Lecture notes (1975), CLNS-321,
    to be published. See also A. Carroll, J. Kogut,
    D. K. Sinclair, and L. Susskind, Cornell preprint
    CLNS-325, to be published.

2. K. Wilson, Physics Reports, proceedings of the
    Paris Conference, to be published.

3. F. J. Dyson, Proc. Roy. Soc. (London) $\underline{A207}$, 395
    (1951); see also Phys. Rev. $\underline{D2}$, 428 (1951).

4. For a review, see K. Wilson, Rev. Mod. Phys. $\underline{47}$,
    773 (1975), or Th. Niemeyer and J. M. M. Leeuwen,
    in C. Domb and M. Green, Eds. <u>Phase Transitions
    and Critical Phenomena</u>, Vol. VI (Academic, to be
    published).

5. L. Kadanoff and A. Houghton, Phys. Rev. $\underline{B11}$, 377
    (1975).

6. T. L. Bell and K. G. Wilson, Phys, Rev. $\underline{B11}$, 3431
    (1975).

7. A. A. Migdal, Zh. E.T.F. $\underline{69}$, 810 (see also 1457)
    (1975).

8. L. Kadanoff (private communication).

DIFFRACTION SCATTERING IN QUANTUM CHROMODYNAMICS[*]

P. Carruthers

Los Alamos Scientific Laboratory

Los Alamos, New Mexico 87544

and

F. Zachariasen[**] (Presented by F. Zachariasen)

California Institute of Technology

Pasadena, California 91109

ABSTRACT

The infrared problem in quantum chromodynamics is studied in order to elucidate aspects of high energy behavior. The integro-differential equation of Cornwall and Tiktopoulos is used to investigate quark-quark scattering in the limit $\lambda \to 0$ (λ being the gluon regulator mass), $s \to \infty$ with t fixed. The solution displays the infrared factors explicitly. When this formula is expanded in power series and the leading ℓog is extracted

---

[*] Research performed under the auspices of the U. S. Energy Research and Development Administration.
[**] On leave from California Institute of Technology, Pasadena, California.

one recovers the perturbation theory calculations to
sixth order. Having argued that the infrared singular
terms in the equation are independent of the renormalization mass M, asymptotic freedom can be used to evaluate
the remainder. Some remarks are made on the scattering
of color-singlet quark clusters with a view towards
solving the physical problem.

I.   INTRODUCTION

What is the asymptotic behavior of diffraction
scattering of hadrons in quantum chromodynamics?[*] The
answer to this question depends critically on understanding the infrared behavior of the theory. First, because
it has been conjectured[1] that confinement of the colored
quarks and gluons out of which physical hadrons are built
is a consequence of the infrared singularities. Second,
because the dominant high energy effects of the scattering
of quarks and gluons are associated with infrared singular terms.

Most prior investigations[2-5] of high energy scattering have been perturbative in nature, both because they
do not deal with hadrons and the bound state problem,
and because they treat the scattering of quarks order by
order. It is most unlikely that either the binding/con-
finement problem or the elimination of infrared singularities can be dealt with in this manner.

---

[*] We do not use the word "Pomeron" here. "Pomeron" has,
in recent years, been used to describe so many conflict-
ing concepts that to employ it now would trigger unnecessary prejudices in the reader's mind. Perhaps the
time has come to retire the word.

However Cornwall and Tiktopoulos[6] have proposed an
integro-differential equation whose solution is conjectured to collect all of the infrared singularities in all
orders of g in the limit $\lambda \to 0$ (λ is a gluon mass cutoff
introduced to render the theory infrared finite); indeed,
when applied to quantum electrodynamics[6,7] the equation
reproduces the well known results of Yennie, et al.[8]
The solutions to this equation, in the fixed angle regime,
suppress the scattering of any particles or clusters of
particles with color by factors which vanish exponentially
in the limit $\lambda \to 0$; in contrast, the suppression factors
are absent in the scattering of colorless objects.
Cornwall and Tiktopoulos see in these facts a hint of
confinement.

If rigorously confirmed, this differential equation
provides a much needed means of transcending perturbation
theory, and in particular a means of dealing with the
infrared aspects of high energy diffraction scattering;
it is this application of it which we wish to exploit
here. We shall discuss quark-quark scattering; the
physical problem of diffraction scattering of hadrons,
which involves bound state effects as well as infrared
effects will not be treated explicitly.

The use of the differential equation to study quark-
quark scattering leads to many advantages over perturbative methods. For example, the extensive cancellations
among individual graphs occurs automatically. Further
one can write for quark-quark scattering a closed form
solution in the case t fixed, $s \to \infty$ whose large s expansion agrees with the $\lambda \to 0$ limit of the leading
logarithm expansion of ref. 3 to 6th order in both the
color flip and non-flip amplitudes.[9] Our explicit formulas
give predictions to all orders which can be confirmed or

denied by further by further evaluations in perturbation
theory.

In Section II we discuss infrared singularities in
QED using the integral equation approach. Section III
treats quark-quark scattering in detail. Our main result
is that the contribution of the infrared singularities
can be exhibited explicitly. Since these are independent
of the arbitrary renormalization point, asymptotic freedom can then be applied (Section IV) to the remainder of
the amplitude to give the complete asymptotic behavior.
However we have not achieved a physical result because
we have not solved the bound state problem. For color-
singlet scattering the "dominant" logarithmic terms
(color octet exchange amplitude for quark-quark scattering) go away so that one has to understand more fully
the non-leading terms. Section IV contains some remarks
on the scattering of clusters.

Some authors imagine that there is a finite gluon
mass λ as a result of spontaneous symmetry violation.
We shall, however, not concern ourselves with this possibility here. Thus the gluon mass λ is inserted only
as an infrared cutoff and we shall always take the limit
as $\lambda \to 0$. Hence the scalar mesons associated with spontaneous symmetry breakdown play no role and can be ignored

The parameters we have to deal with, therefore, are
as follows: the invariant energy and momentum transfers
s, t, and u; the infrared cutoff λ; (possibly) a quark
mass m, and a mass M at which the renormalization is
carried out. (When we discuss clusters, additional
parameters, such as cluster masses, may appear as well.)
Associated with M there is a coupling constant g_M; however
a physical amplitude cannot depend separately on M and
g_M, since M can be chosen arbitrarily. In the absence

of λ, and for large M, only the renormalization group invariant combination $\mu^2 = Me^{2^{1/bg_M^2}}$ can actually occur, where $\beta(g) = bg^3 + \ldots$, as usual, defines b.

This set of parameters is large enough so that it will help, at this point, to lay out clearly the limits we shall look at.

We always take $\lambda \to 0$ first; that is, λ^2/s, λ^2/t, λ^2/μ^2, (and λ^2/m^2 if $m \neq 0$), all vanish. We next take s and $- u$ to infinity. Specifically, $s/\mu^2 \to \infty$ (and $s/m^2 \to \infty$ if $m \neq 0$) and similarly for $- u$. We shall be primarily interested in the fixed t case, so that $s/t \to \infty$ as well, though to make connections with previous work we shall also look at the fixed angle regime, where $s/t \sim 1$. In all of the above we shall distinguish the zero quark mass (m=0) situation from the finite quark mass (m≠0) case, since the results are slightly different in the two cases.

II. QED

We shall first go through the sequence of limits in conventional quantum electrodynamics, since here the answers are generally well known,[8] and can be used to normalize our techniques.

Let two quarks of momenta p_1 and p_2 scatter from each other and end up with momenta p_1' and p_2'. Then $s = 2p_1 \cdot p_2 + 2m^2$, $t = -2p_1' \cdot p_1 + 2m^2$ and $u = -2p_1' \cdot p_2 + 2m^2$, with $s+t+u = 4m^2$.

Our starting point will be the differential equation introduced by Cornwall and Tiktopoulos,[6] for the purpose of extracting the infrared singularities of scattering amplitudes. They write, as $\lambda \to 0$, for QED,

$$\lambda\frac{\partial}{\partial\lambda}T(p_1\cdots p_n)=\lambda\frac{\partial}{\partial\lambda}T_B(p_1\cdots p_n)+\tfrac{1}{2}\!\int\!\frac{d^4k}{(2\pi)^4}\sum_{i,j}\eta_i\eta_j$$

$$\times\qquad K(\eta_1 p_i,\eta_j p_j,k;\lambda)\qquad\qquad (2.1)$$

$$\times\qquad T(p_1,\ldots,p_1-\eta_i k,\ldots,p_j+\eta_j k,\ldots,p_n)\qquad,$$

where i and j run over all particles in the process, where $\eta_i=+1(-1)$ for out (in) going particles, where

$$K(p,p',k;\lambda)=-8ig^2\frac{\lambda^2}{(k^2-\lambda^2)^2}\frac{p\cdot p'}{(k^2-2pk)(k^2+2p'k)}\quad.\qquad (2.2)$$

and where T_B stands for the Born approximation. An important comment is in order here. Equation (2.2) contains a coupling constant g, and we must ask which g this is. In QED, g is known to be the physical coupling constant, renormalized with all particles on their mass shell. (This is of course plausible since the infrared singularity is associated with particles which are near their mass shell). In non-Abelian gauge theories (to which we shall turn in the following section) the analogue could be g_λ, the coupling constant renormalized with the quarks on their mass shell but with the gluon renormalized at λ, the infrared cutoff. It could also be g_k^2, renormalized at the variable gluon mass in the integral. One cannot decide which g occurs by studying only leading logarithm approximations to the perturbation series, since to this order the series is not changed by a shift in renormalization point. It is necessary to look at non-leading logarithms, and in fact a study of non-leading renormalization corrections to eq. (2.1) encourages the belief that g_k^2 is the proper coupling to use. We should also remark that in contrast to QED, in non-Abelian gauge theories g_k^2 probably does not approach a finite limit as $k^2 \to 0$, but is likely to be singular instead.

If we apply eq. (2.1) to the special case of electron-electron scattering, then we obtain, upon neglect of the diagonal terms in (2.1),

$$\lambda \frac{\partial}{\partial\lambda} T(P,Q,R;\lambda) = \lambda \frac{\partial}{\partial\lambda} T_B(P,Q,R;\lambda) + \int \frac{d^4k}{(2\pi)^4}$$

$$\{[K(p_1',p_2',k;\lambda)T(P,Q-k,R-k;\lambda)$$

$$+ K(-p_1,-p_2,k;\lambda)T(P,Q-k,R+k;\lambda)]$$

$$-[K(p_1',-p_1,k;\lambda)T(P-k,Q,R-k;\lambda)$$

$$+ K(p_2',-p_2,k;\lambda)T(P-k,Q,R+k;\lambda)] \tag{2.3}$$

$$-[K(p_1',-p_2,k;\lambda)T(P-k,Q-k,R;\lambda)$$

$$+ K(-p_2',p_1,k;\lambda)T(P+k,Q-k,R;\lambda)]\} .$$

Here we have chosen to use $P = p_1+p_2$, $Q=p_1'-p_1$, and $R = p_1'-p_2$ as independent variables, with $p^2 = s$, $Q^2 = t$ and $R^2 = u$. This equation is supposed to be valid, as mentioned above, in the limit where the infrared cutoff is small compared to all other mass parameters (namely s, t, u, the renormalization group invariant mass μ^2, etc.). If in addition we let s and u approach infinity, relative to μ^2 (and m^2, if $m\neq0$), then we may neglect k compared to P and R in eq. (2.3); thus

$$\lambda \frac{\partial}{\partial \lambda} T(P,Q,R;\lambda) = \lambda \frac{\partial}{\partial \lambda} T_B(P,Q,R;\lambda) +$$

$$\int \frac{d^4 k}{(2\pi)^4} [K_s(k,\lambda) - K_u(k,\lambda)] \, T(P,Q-k,R;\lambda) - 2\tilde{B}(t,\lambda) T(PQR;\lambda). \quad (2.4)$$

Here we use the shorthand

$$K_s(k,\lambda) \equiv K(p_1',p_2',k;\lambda) + K(-p_1,-p_2,k;\lambda), \quad (2.5)$$

with similar expressions for K_t and K_u, and we define[*]

$$\tilde{B}(2p' \cdot p) \equiv \int \frac{d^4 k}{(2\pi)^4} \, K(p,p',k;\lambda)$$

$$= \int \frac{d^4 k}{(2\pi)^4} \frac{\lambda^2}{(k^2 - \lambda^2)^2} \frac{-8ig^2 p' \cdot p}{(k^2 - 2p \cdot k)(k^2 + 2p' \cdot k)} \quad . \quad (2.6)$$

We have not been able to write down in closed form the general solution to eq. (2.4); we can, however, write the solution for $s \to \infty$ in the fixed angle regime, where k can be neglected compared to Q as well as to P and R.

In the fixed angle regime, one can neglect the Born term, and (2.4) simplifies to

$$\lambda \frac{\partial}{\partial \lambda} T = 2[\tilde{B}(s,\lambda) - \tilde{B}(t,\lambda) - \tilde{B}(u,\lambda)] T \quad (2.7)$$

with the solution

$$T(P,Q,R;\lambda) = \exp\left(\int \frac{d\lambda'}{\lambda'} 2[\tilde{B}(s,\lambda') - \tilde{B}(t,\lambda') - \tilde{B}(u,\lambda')]\right) \bar{T}(PQR). \quad (2.8)$$

---

[*] Our $\tilde{B}$ is related to the function B defined by Cornwall and Tiktopoulos (ref. 6) through $\tilde{B} = \lambda \partial B/\partial \lambda$.

As $s \to \infty$, with s, t and u all of the same order, and when
m = 0

$$\tilde{B}(s,\lambda) \to -\frac{g^2}{4\pi^2} [\log s/\lambda^2 - i\pi] ,$$

$$\tilde{B}(u,\lambda) \to -\frac{g^2}{4\pi^2} [\log s/\lambda^2 + \log \alpha_u] ,$$

$$\tilde{B}(t,\lambda) \to -\frac{g^2}{4\pi^2} [\log s/\lambda^2 + \log \alpha_t] , \qquad (2.9)$$

where α_u, $\alpha_t = \frac{1}{2}(1 \pm \cos\Theta)$ plus terms that vanish with λ.
When $m \neq 0$, the same formulae hold with $\log s/\lambda^2$ replaced
by $\log s/m^2$. Hence (2.8) yields, when m = 0

$$T(P,Q,R;\lambda) = \exp\left\{-\frac{g^2}{8\pi^2} \log^2 s/\lambda^2\right.$$
$$\left.-i \frac{g^2}{4\pi} \log s/\lambda^2 - \frac{g^2}{2\pi^2} \log \frac{\sin\Theta}{2} \log s/\lambda^2\right\} \bar{T}(P,Q,R)$$
$$(2.10)$$

and when $m \neq 0$ it yields

$$T(P,Q,R;\lambda) = \exp\left\{-\frac{g^2}{4\pi^2} \log s/\lambda^2 \log s/m^2\right.$$
$$\left.-i \frac{g^2}{4\pi} \log s/\lambda^2 - \frac{g^2}{2\pi^2} \log \frac{\sin\Theta}{2} \log s/\lambda^2\right\} \bar{T}(P,Q,R) ,$$
$$(2.11)$$

where $\bar{T}(P,Q,R)$ is independent of λ. These results display all of the infrared singularities of the amplitude
in the fixed angle regime.

   In the fixed t regime the analysis is more complicated since close inspection shows that none of the terms
in (2.4) can be dropped as $s \to \infty$. In fact the situation in
QED is more complicated than in QCD. In the latter theory,
one has $k_s + k_u$ rather than $k_s - k_u$ in the significant
integral. As a consequence the $\tilde{B}(t)T$ term can be dropped

(cf. Eq. (3.14)) and the resulting equation easily solved
in configuration space. As a consequence of these small
differences, which are directly traceable to the group
properties of QCD, the asymptotic behavior of the two
theories is quite different. We plan to discuss this
question elsewhere.

III. COLOR

When we introduce color, the basic equation (2.1)
changes only in that the gluon coupling to the ith quark
picks up a color operator factor $\vec{t}_i$. We are dealing with
an amplitude with only external quark lines; therefore
glue-glue interactions (either cubic or quartic) do not
explicitly contribute to the infrared singular part.[*]

Equation (2.1) becomes

$$\lambda\frac{\partial}{\partial\lambda} T = \lambda\frac{\partial}{\partial\lambda} T_B + \int\frac{d^4k}{(2\pi)^4} \sum_{ij} K(p_i\eta_i, p_j\eta_j, k, \lambda)\eta_i\eta_i$$
$$(\vec{t}_i\cdot\vec{t}_j \ T(p_1,\ldots,p_i-\eta_i k,\ldots,p_j+\eta_j k,\ldots,p_n) \ )+ \tag{3.1}$$

and the ( )$_+$ symbol is designed to order the $\vec{t}$ operators
so that those associated with incoming quarks stand to
the right of T, while those attached to outgoing quarks
stand to the left. Thus (2.3) is replaced by

---

[*] They do, of course, contribute. However, as in QED
their effect is to cancel the difference between the
true Feynman diagrams with gluons coupled to external
quarks and the effective Feynman diagrams, employing
spinless quark propagators and scalar coupling of quarks
to gluons, used in eq. (3.1).

$$\lambda \frac{\partial}{\partial \lambda} \, T(P,Q,R;\lambda) \; = \; \lambda \frac{\partial}{\partial \lambda} \, T_B(P,Q,R;\lambda) + \int \frac{d^4 k}{(2\pi)^4}$$

$$\{ [K(p_1',p_2',k,\lambda)\vec{t}_1' \cdot \vec{t}_2' \; T(P,Q-k,R-k;\lambda)$$

$$+K(-p_1,-p_2,k,\lambda)T(P,Q-k,R+k)\vec{t}_1 \cdot \vec{t}_2]$$

$$-[K(p_1',-p_1,k,\lambda)\vec{t}_1' \cdot T(P-k,Q,R-k)\vec{t}_1$$

$$+K(p_2',-p_2,k,\lambda)\vec{t}_2' \cdot T(P-k,Q,R+k)\vec{t}_2]$$

$$-[K(p_1',-p_2,k,\lambda)\vec{t}_1' \cdot T(P-k,Q-k,R)\vec{t}_2$$

$$+K(-p_2',p_1,k\lambda)\vec{t}_2' \; T(P+k,Q-k,R)\vec{t}_1] \} \; . \qquad (3.2)$$

Again we approximate (3.2) by assuming s and u to be large compared to m^2, if any, and to μ^2, so that P and R dominate k. Then, as before, there are two regimes of interest, fixed angle and fixed t. We look first at fixed angle.

A. The Fixed Angle Regime

For elastic scattering we may set $\vec{t}_1'=\vec{t}_1$ and $\vec{t}_2'=\vec{t}_2$. The analogue of eq. (2.12) is then

$$\lambda \frac{\partial}{\partial \lambda} \, T=\tilde{B}(s,\lambda)(\vec{t}_1 \cdot \vec{t}_2 \; T+T\vec{t}_1 \cdot \vec{t}_2)$$

$$-\tilde{B}(t,\lambda)(\vec{t}_1 \cdot T\vec{t}_1+\vec{t}_2 \cdot T\vec{t}_2)$$

$$-\tilde{B}(u,\lambda)(\vec{t}_1 \cdot T\vec{t}_2+\vec{t}_1 \cdot T\vec{t}_1) \; . \qquad (3.3)$$

For SU(2) as the color group $\vec{t}=\frac{1}{2}\vec{\tau}$ and we decompose T into color non-flip and color flip amplitudes according to

$$T = T_0+\vec{\tau}_1 \cdot \vec{\tau}_2 T_1 \quad . \qquad (3.4)$$

Equation (3.3) becomes a pair of coupled differential equations for T_o and T_1:

$$\lambda \frac{\partial T_o}{\partial \lambda} = \frac{3}{2}[-\tilde{B}(t)T_o+(\tilde{B}(s)-\tilde{B}(u))T_1] \ ,$$

$$\lambda \frac{\partial T_1}{\partial \lambda} = \frac{1}{2}[(\tilde{B}(s)-\tilde{B}(u))T_o-(2\tilde{B}(s)+2\tilde{B}(u)-\tilde{B}(t))T_1] \ . \tag{3.5}$$

(i) Zero Quark Mass.

First look at the m=0 case. Using the asymptotic forms (2.9)[*] we write the asymptotic form of equations (3.5) in the form

$$-\frac{\partial T_o}{\partial x} = \alpha x T_o+3i\beta \ T_1 \ ,$$

$$-\frac{\partial T_i}{\partial x} = i\beta T_o+\alpha x T_1 \ , \tag{3.6}$$

where $x = \ln s/\lambda^2$, $\alpha=3g^2/16\pi^2$, $\beta=g^2/16\pi^2(\pi-i\log\alpha_u)$. The solutions to these equations are

$$T_o=-i\sqrt{3} \ \sin(\sqrt{3}\beta x+\phi) \ e^{-\frac{1}{2}\alpha x^2}\bar{T}_1 \ ,$$

$$T_1= \cos(\sqrt{3}\beta x+\phi) \ e^{-\frac{1}{2}\alpha x^2}\bar{T}_1 \ , \tag{3.7}$$

where the integration constants ϕ and $\bar{T}_1$ are independent of λ. In the SU(3) case we write $T=T_o+\vec{\lambda}_1 \cdot \vec{\lambda}_2 T_1$ and find in place of (3.5)

---

[*] Note that these forms ignore the possible k dependence in g, which in the non-Abelian case may well be present.

$$\lambda \frac{\partial}{\partial \lambda} T_o = \frac{8}{3}[-\tilde{B}(t)T_o + \frac{2}{3}(\tilde{B}(s)-\tilde{B}(u))T_1] \quad,$$

$$\lambda \frac{\partial}{\partial \lambda} T_1 = \frac{1}{2}\{(\tilde{B}(s)-\tilde{B}(u))T_o - \frac{1}{3}(2\tilde{B}(s)+7\tilde{B}(u)-\tilde{B}(t))T_1\}. \quad (3.8)$$

Asymptotically eqs. (3.8) have the form (3.6) with 3 replaced by 32/9 and with differing α and β: $\alpha = g^2/3\pi^2$, $\beta = g^2/16\pi^2(\pi-i\ell n\alpha_u)$. Hence the solutions have the same form as (3.8) with $\sqrt{3} \rightarrow \sqrt{32/9}$.

Note that while the dominant logarithms cancel in the difference $\tilde{B}(s)-\tilde{B}(u)$ we are left with some angle dependence via the factor $\log\alpha_u$. Since T_1 dominates T_o by one power of logs in the leading log approximation and since in this approximation the cosine in (3.7) is unity, this angle dependence appears only in subdominant terms. More precisely we consider the limit in which $g^2\log^2 s$ is held finite while $g \rightarrow 0$, $\log s \rightarrow \infty$. Then the solutions (3.7) become

$$T_o = -i\frac{3g^2}{16\pi^2}(\pi-i\log\alpha_u)\log s/\lambda^2 \ e^{-\frac{3g^2}{32\pi^2}\log^2 s/\lambda^2} \ \bar{T}_1,$$
$$(3.9)$$
$$T_1 = e^{-\frac{3g^2}{32\pi^2}\log^2 s/\lambda^2} \ \bar{T}_1 \quad .$$

Thus the perturbation series for T_o (at fixed angle) in the leading log approximation is

$$T_o = -3i\frac{g^2}{8\pi}\frac{s}{t}\log s/\lambda^2 \ (1-\frac{3g^2}{32\pi^2}\log^2 s/\lambda^2+...) \ (3.10)$$

since $\bar{T}_1 = 2g^2 s/t$ to lowest order in g.

Note that this limit also results from ignoring the contribution of T_o in the differential equation for T_1; this is evidently acceptable order by order in perturbation theory ((i.e.), in the leading log limit). If this is done, the equation can be formally integrated even if g does depend on k^2:

$$T_1 \cong e^{-\frac{3}{2} \int \frac{d\lambda}{\lambda} \tilde{B}(s,\lambda)} \; \bar{T}_1 \; , \qquad (3.11)$$

where $\tilde{B}(s,\lambda) \propto g^2$ (cf. Eqs. 2.9).

The same approximation in the T_o equation expresses T_o simply as an integral over T_1:

$$T_o = -\frac{3}{2} \int \frac{d\lambda}{\lambda} \; (\tilde{B}(s,\lambda) - \tilde{B}(u,\lambda)) T_1(\lambda). \qquad (3.12)$$

(ii) Non-Zero Quark Mass:

The situation when $m \neq 0$ is slightly simpler. Now (if we again ignore any k dependence in g) the functions $\tilde{B}$ are all independent of λ so that eqs. (3.5) become simply a pair of coupled linear equations with constant coefficients. We content ourselves with writing down the analogues of eqs. (3.9), which arise from neglecting non-leading logarithms in (3.5). We find

$$T_1 = e^{-\frac{3g^2}{16\pi^2} \log s/\lambda^2 \; \log s/m^2} \; \bar{T}_1 \qquad (3.13)$$

$$T_o = 4i(\pi - i\log\alpha_u)(e^{-\frac{3g^2}{16\pi^2} \log s/\lambda^2 \; \log s/m^2} - 1)\bar{T}_1.$$

B.   The Fixed t Regime

Next we turn to the fixed t regime. Now (3.2) become the coupled system (again with SU(2) as the color group)

$$\lambda\frac{\partial}{\partial\lambda}T_o(P,Q,R;\lambda) = \frac{3}{4}\int\frac{d^4k}{(2\pi)^4}[K_s(k,\lambda) - K_u(k,\lambda)]T_1(P,Q-k,R;\lambda)$$

$$\lambda\frac{\partial}{\partial\lambda}T_1(P,Q,R;\lambda) = \lambda\frac{\partial}{\partial\lambda}T_B(P,Q,R;\lambda) + \frac{1}{4}\int\frac{d^4k}{(2\pi)^4}[K_s(k,\lambda) - K_u(k,\lambda)]T_o$$

$$\times(P,Q-k,R;\lambda) - \frac{1}{2}\int\frac{d^4k}{(2\pi)^4}[K_s(k,\lambda) + K_u(k,\lambda)]T_1(P,Q-k,R;\lambda).$$

$$(3.14)$$

Fourier transforming on Q simplifies these:

$$\lambda \frac{\partial}{\partial \lambda} T_o(P,x,R;\lambda) = \frac{3}{4} K_-(x,\lambda) T_1(P,x,R;\lambda) ,$$

$$\lambda \frac{\partial}{\partial \lambda} T_1(P,x,R;\lambda) = \lambda \frac{\partial}{\partial \lambda} T_B(P,x,R;\lambda) + \frac{3}{4} K_-(x,\lambda) T_o(P,x,R;\lambda) ,$$

$$-\frac{1}{2} K_+(x,\lambda) T_1(P,x,R;\lambda) , \qquad (3.15)$$

where $K_\pm$ are $K_s \pm K_u$.

The K_+ term will dominate over the K_- term order by order in the second of these, in analogy to the fixed angle situation. Thus the result for this case is

$$T_1(P,Q,R;\lambda) = \int d^4x \ e^{iQ \cdot x} \int_\infty^\lambda d\lambda' \frac{\partial}{\partial \lambda'} T_B(P,x,R;\lambda')$$

$$\times \exp\left(-\frac{1}{2} \int_{\lambda'}^\lambda \frac{d\lambda''}{\lambda''} \int \frac{d^4k}{(2\pi)^4} e^{-ikx} K_+(k,\lambda'')\right) . \qquad (3.16)$$

Our integration by parts, the similarity of eq. (3.16) to the Rytov, or super-eikonal approximation[10,11] becomes evident. T_o is found by direct integration:

$$\qquad\qquad\qquad\qquad\qquad\qquad\qquad\qquad\qquad (3.17)$$

$$T_o(P,Q,R;\lambda) = \frac{3}{4} \int dx \ e^{iQx} \int_\infty^\lambda \frac{d\lambda'}{\lambda'} K_-(x,\lambda') T_1(P,x,R;\lambda') .$$

To enhance our belief in the basic equation, it is of some value to compare these solutions with leading log perturbation calculations. To this end, we expand the expression for T_1 in powers of g^2, as follows:

$$T_1(P,Q,R;\lambda) = 4sg_t^2 \sum_{k=0}^{\infty} \frac{(-1)^n}{n!} \int_\infty^\lambda \lambda' d\lambda' \qquad (3.18)$$

$$\prod_{i=1}^n \int_{\lambda'}^\lambda \frac{d\lambda_i}{\lambda_i} \int \frac{d^4k_i}{(2\pi)^4} K(k_i,\lambda_i) \left[\frac{1}{(Q-k_1-\ldots-k_n)^2 - \lambda'^2}\right]^2 .$$

To reproduce the perturbation series, in the leading log approximation, we first ignore the variation in g, and evaluate the integral on $d\lambda'$. It is then directly obvious, just as in the QED case, when one writes out K_+ explicitly, that the order g^4 term is exact, and precisely reproduces the box graphs. In each higher order, the leading logs and leading log t/λ^2 can be easily extracted, and one obtains, when $m \neq 0$

$$T_1 = \frac{2sg^2}{t} \sum \frac{1}{n!} \left(-\frac{g^2}{4\pi^2}\right)^n (\log s/m^2 \; \log - t/\lambda^2)^n = \frac{2sg^2}{t} e^{-\frac{g^2}{4\pi^2} \log s/m^2 \; \log t/\lambda^2}$$

(3.19)

in complete agreement with conventional perturbation theory. The Reggeization is to be expected on the basis of previous investigations.[12,13]

Similar operations may be performed on the expression for T_o, and one obtains

$$T_o = -\frac{3i}{8\pi} g^4 \frac{s}{t} \log \frac{-t}{\lambda^2} \sum \frac{1}{(n+1)!} \left(-\frac{g^2}{4\pi^2}\right)^n \left(\log s/m^2 \; \log t/\lambda^2\right)^n$$

$$= \frac{3i\pi}{2} g^2 \frac{s}{t} \left(\frac{e^{-\frac{g^2}{4\pi^2} \log s/m^2 \; \log - t/\lambda^2} - 1}{\log s/m^2}\right).$$

(3.20)

This expression also agrees with conventional perturbation through order g^6, which is as much as has been checked so far.

When $m = 0$, nothing much changes. It is easily seen that in (3.19) and (3.20) one simply replaces $\log s/m$ by $\log s/\lambda^2$.

Equation (3.20) (with $\log s/m^2 \to \log s/\lambda^2$ if $m = 0$) gives the leading logs infrared singular behavior of the

quark-quark scattering amplitude in the $s \to \infty$, t fixed
limit. If one wishes to describe this amplitude in terms
of j-plane singularities, one would say that it consists
of two logarithmic j-plane branch cuts, one fixed at
j=1 and one located at the same place as the Regge pole
describing the isospin flip amplitude T_1. More of the
infrared structure is contained in the expressions (3.16)
and (3.17) (though even these do not encompass the entire
λ dependence of the amplitude - for that, one has to return
to eq. (3.2); however a j-plane description of these is
difficult and seems not particularly useful.

IV. CLUSTERS, ASYMPTOTIC FREEDOM, AND CONCLUSIONS

What happens if we scatter two clusters of free
quarks-that is, groups of quarks having small relative
momenta, from each other. Again we return to the fundamental differential equation (3.1).

We first argue that gluon exchange between two quarks
in the same cluster can be neglected, to leading order,
because no large logarithm can result from the function
K_{ij} when the relative momentum of the two quarks is small.
we next argue, for the same reason, that K_{ij} is the same,
to leading order, for any pair of quarks in two different
clusters, when the clusters are separated by a large
momentum difference (P and R for fixed t, P Q and R for
fixed angle). We finally argue that any exchange between
quarks in clusters separated by a small momentum difference (Q for fixed t) can be ignored, as in the simple
quark-quark scattering case.

From these arguments, we conclude that (3.2) is still
valid, where P Q and R refer to cluster momenta, but where
each $\vec{t}$ is replaced by the total color operator of the
cluster:

$$\vec{t} \to \sum_{i=1}^{n_c} \vec{t}_i$$

for n_c particles with colors t_i, $i=1..n_c$, making up the
cluster. So all our previous results should hold for
cluster scattering: the IR singularities for clusters can
be extracted as before.

Now what if the clusters are all colorless? Then
the cluster color operator $\sum_{i=1}^{n_c} \vec{t}_i$ is identically zero,
when acting on any of the external clusters. Thus, in
both regimes, all leading infrared singularities dis-
appear. This does not, of course, mean that T has no
infrared singularities; there surely are singularities
involving, for example, cluster masses divided by λ^2.
It simply means that the leading log λ in the $s\to\infty$ limit
all cancel out. The hope must be that this non-dominant
kind of infrared behavior is associated with the binding
of the clusters to produce physical hadrons.

If this hope is realized, it suggests that one
approach to hadronic scattering may be to start with colorless clusters of free quarks, to observe that the leading
(as $s\to\infty$) infrared singularities cancel,to separate the
non-leading infrared singularities into bound state wave
functions, and to identify the physical diffractive
asymptotic behavior with the remaining energy dependence.

It also means that in perturbation theory, hadronic
scattering will arise from non-leading (as $s\to\infty$) and non-
infrared singular contributions in each graph. Thus a
perturbative approach which focuses on leading logs in
the scattering of colorful objects only will have a very
hard time seeing any signals associated with physical
effects. It will be necessary to peel away, in each order
the infrared singular terms and to associate the remaining

terms with true scattering.

We could imagine trying to set up such a calculation as follows. We begin with the basic equation (3.1), but now, in going on to (3.14) we retain the terms we neglected when we replaced P±k and R±k by P and R. Let us call those terms X. Then (3.14) is modified to read (again for SU(2))

$$\lambda \frac{\partial}{\partial \lambda} T_o = \frac{3}{4} \int dk \ K_- T_1 - \frac{3}{2} \tilde{B} \ T_o + X_o \quad , \qquad (4.1)$$

$$\lambda \frac{\partial}{\partial \lambda} T_1 = \lambda \frac{\partial}{\partial \lambda} T_B + \frac{1}{4} \int dk \ K_- T_o - \frac{1}{2} \int dk \ K_+ T_1 + \frac{1}{2} \tilde{B} \ T_1 + X_1 .$$

In writing this we have also reintroduced the $\tilde{B}(t) \times T(P,Q,R;\lambda)$ terms neglected in (3.14). Finally, let us assume that the coupling constant g in $K_\pm$ and $\tilde{B}$ is g_k^2 and that in T_B is g_Q^2.

The functions X_o and X_1 depend, of course, on T; they are not explicitlyly known. But they are infrared finite and hence independent of λ.

Since the functions X do not depend on the IR cutoff, they can be functions only of s/M^2, T/M^2, m^2/M^2 and g_M, where M is the UV renormalization point and g_M is the associated coupling constant. But T does not depend on M; nor do the IR singular terms. Hence under a change in M, X_o and X_1 pick up no anomalous dimension; thus we would guess that

$$(M \frac{\partial}{\partial M} + \beta \frac{\partial}{\partial g}) \ X_{o,1} = 0 . \qquad (4.2)$$

It therefore might suffice to calculate X_o and X_1 in lowest order perturbation theory, which will be order g^4.

In this way everything in (4.1) is known, and it becomes an equation for the amplitude T. Furthermore X_o and X_1 introduce s dependence coupled to the mass μ, not λ. This s dependence is, of course, non leading; as we have seen, the leading s dependence is always associated with the infrared behavior. Nevertheless the nor infrared non-dominant μ-associated s dependence may well be connected with physical high energy behavior. One can conceive of the solution to (4.1) permitting the identification of both the unphysical infrared high energy behavior and the perhaps physical variation with s/μ^2. One might even conceive of the possibility of identifying the hadronic cross section by guessing that (when m=0 at least) it will have the form $\sigma(s) = \frac{1}{\mu^2} f(s/\mu^2)$.

REFERENCES

1. See, e.g., S. Weinberg, Phys. Rev. Letters $\underline{31}$, 493 (1973), and D. Gros and F. Wilczek, Phys. Rev. Letters $\underline{30}$, 1343 (1973).

2. L. Tyburski (Univ. of Illinois preprint, to be published).

3. B. M. McCoy and T. T. Wu, Phys, Rev. $\underline{D12}$, 3257 (1975) and to be published.

4. C. Y. Lo and H. Cheng (M. I. T. preprint, to be published).

5. E. Poggio and H. R. Quinn, Phys. Rev. $\underline{D12}$, 3279 (1975).

6. J. M. Cornwall and G. Tiktopoulos, Phys. Rev. Letters and UCLA preprint (to be published).

7. C. P. Korthals Altes and E. de Rafael (to be published).

8. D. R. Yennie, S. Frautschi and H. Suura, Ann. of Phys. $\underline{13}$, 379 (1961).

9. The perturbation results were given in the case of a local SU_2 gauge symmetry. In our approach it is elementary to change from SU_2 to SU_3.

10. For a description of the Rytov method, see L. Chernov "Wave Propagation in a Random Medium" McGraw-Hill Book Co., New York, N.Y. 1960.

11. W. H. Munk and F. Zachariasen, JASA, to be published.

12. The results of refs. 2-4 were given in terms of certain integrals $K_i(t)$. We have evaluated these integrals for small λ in order to compare with our equations.

13. M. Grisaru, H. J. Schnitzer, and H. Tsao, Phys. Rev. Letters 30, 811 (1973). It is worth remarking that in the limit $\lambda \to 0$, the vector meson no longer lies on the Regge trajectory so generated. When $\lambda \to 0$, the trajectory function $\alpha(t)$, far from being one, is singular at $t = 0$.

HIGH-ENERGY BEHAVIOR OF TWO-BODY ELASTIC AMPLITUDES IN GAUGE FIELD THEORIES

Barry M. McCoy[*]

Institute for Theoretical Physics

State University of New York at Stony Brook

Stony Brook, New York 11794

and

Tai Tsun Wu[†] (Presented by Tai Tsun Wu)

Gordon McKay Laboratory

Harvard University

Cambridge, Massachusetts 02138

In this talk, we discuss the high-energy behavior of two-body elastic amplitudes in gauge field theories. We shall be concerned only with the limit:

$$s \to \infty, \tag{1}$$

with t fixed at a physical value.

---

[*] Alfred P. Sloan Fellow. Work supported in part by the National Science Foundation under Grant No. DMR 73-7565 A01

[†] Work supported in part by the U.S. Energy Research and Development Administration under Contract No. E(11-1)-3277.

We shall discuss the following cases
1. Abelian gauge theory (electrodynamics)
    A. Pomeron exchange
    B. Quantum number exchange
2. Non-Abelian gauge theory (Yang-Mills theory[1])

Before going into the details, let us compare the present limit (1) with that discussed by Zachariasen a few minutes ago, and by Cornwall yesterday. Here we avoid entirely the problems of infrared divergence due to massless particles. This is accomplished in the case of quantum electrodynamics by artificially introducing a mass in the photon propagator. In the case of the Yang-Mills theory[1], the mass of the gauge particle is introduced through the Higgs' mechanism.[2] For the isospin group SU_2. A doublet of complex scalar particles is sufficient to render all particles massive.

In order to study the high-energy behavior of these field theories, we follow the so-called procedure of "summing the leading terms". We emphasize that the correctness of this procedure of summing the leading terms is a pure assumption. Because of the important role played by this procedure, let us discuss it in some detail.

The rules of the game are the following:
(a) For each order of perturbation theory, calculate the leading behavior of the scattering amplitude in the limit(1).
(b) If, for a higher order of perturbation theory, the leading term is of the same order of magnitude as that for a lower order; then the term from the higher order is discarded.
(c) Sum the remaining terms.

As an example, consider the case of electron-electron elastic scattering in case 1A. Here the orders of magnitudes for contributions from various orders of perturbation theory are as follows:

$$
\begin{array}{ll}
\text{2nd order} & 0 \\
\text{4th order} & s \\
\text{6th order} & s \\
\text{8th order} & s \ln s \\
\text{10th order} & s \ln s \\
\text{12th order} & s(\ln s)^2 \\
\text{14th order} & s(\ln s)^2 \\
\text{16th order} & s(\ln s)^3
\end{array}
$$

[The result for second order is 0 because the Pomeron is defined to be +1 under charge conjugation C.] By (b) above, we neglect the contributions from 6th., 10th., 14th.,... orders of perturbation. We then sum up the leading terms from 4th., 8th., 12th., 16th.,... orders of perturbation.

As already seen from this example, when applying this procedure of summing the leading terms, we deal with one scattering amplitude at a time. In this example we treat the cases C=+1 and C=-1 separately. As we shall see, this point is most important in case 1B, where the amplitudes for positive signature and negative signature must be studied separately. The reason is that, for each order, the negative-signature amplitude is smaller than the positive-signature amplitude by a factor ln s. However, the sum of leading terms for the negative-signature amplitude turns out to be larger than that for the positive-signature amplitude.

In statistical mechanics, this procedure of summing the leading terms has had a great deal of success[3]. However, even in this somewhat simpler context, there is

at least one example of failure.[4] You must decide for
yourself whether this example is pathological or not.

I. ABELIAN GAUGE THEORY (ELECTRODYNAMICS)

A. Pomeron Exchange

These cases of Pomeron exchange in quantum electrodynamics and scalar electrodynamics have been known for
a number of years. Let us describe them very briefly.
The leading terms come from the tower diagrams, and the
sum of these leading terms yields[5], give or take a few
ln s, is

$$is^{1+A\alpha^2} , \tag{2}$$

where

$$A= \begin{cases} \dfrac{11\pi}{32} & \text{for quantum electrodynamics} \\[3em] \dfrac{5\pi}{32} & \text{for scalar electrodynamics.} \end{cases}$$

Since A>0, we say that the bare Pomeron is above 1. By
iterating in the direct, or s, channel,[6] the Froissart
bound[7] is saturated,

$$\sigma_{total} \sim (ln\ s)^2. \tag{3}$$

This is the <u>rising</u> <u>total</u> <u>cross</u> <u>section</u>.[6]

Let us now review briefly the phenomenological σ_{total}
developed by combining these results from field theory
with the optical model.

(T1) From old data (i.e., data from the Brookhaven
AGS, CERN-PS, and especially Serpukhov), estimate the
increase of proton-proton total cross section over the

energy range of CERN-ISR.[8,11]

(E1) Measurement at CERN-ISR shows an increase about 50% larger than the phenomelogical estimate.[9]

(T2) From old data and this result from CERN-ISR on the proton-proton total cross section, predict total cross sections for π^+p, π^-p, K^+p, K^-p, and $\bar{p}p$.[10,11]

(E2) Accurate measurements at Fermilab on the total cross sections for π^+p, π^-p, K^+p, and K^-p give excellent agreement with the phenomenological predictions.[12] The poorer agreement for the $\bar{p}p$ case can be blamed on the inaccuracy of old data in this case.

Extension of the phenomenological predictions to higher energies has been carried out.[11]

B. Quantum Number Exchange

The high-energy behavior of fermion exchange in quantum electrodynamics was studied first in pre-historic times by Gell-Mann, Goldberger, Low, Marx and our chairman today.[13] For the two processes $e+\bar{e}\rightarrow\gamma+\gamma$ and $e+\gamma\rightarrow\gamma+e$, there is a positive-signature amplitude and a negative-signature amplitude.

Let us consider the case of the sixth order in some detail. As first pointed out by Federbush, the leading term for the positive-signature amplitude in the sixth order comes from the three diagrams of Fig. 1. Omitting a factor of $s^{\frac{1}{2}}$ which may or may not be present depending on the definition of the amplitude, this positive-signature amplitude in sixth order is of the order of $(\ln s)^2$.

In the pioneering work of Gell-Mann etc.,[13] these same three diagrams are used to study the negative-signature amplitude. In this way, they obtained a Regge trajectory that is higher for larger values of physical momentum transfer. This is impossible.

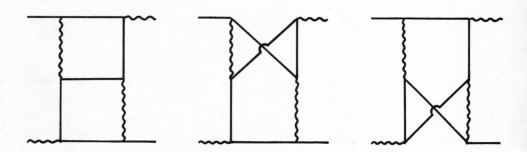

Fig. 1 The three sixth order diagrams that contribute to
the leading terms of both the positive and the negative
signature amplitudes for fermion exchange in quantum
electrodunamics at high energies. (The s-channel is
from left to right, while the t-channel is up-down.)

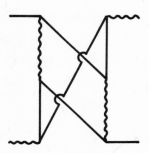

Fig. 2 The additional diagram that contributes to the
leading term of the negative-signature amplitude but not
to that of the positive-signature amplitude.

The resolution of this difficulty lies in the observation that for the negative-signature amplitude in the sixth order, a fourth diagram is needed. This is shown in Fig. 2. This diagram contributes only to the leading term of the negative-signature amplitude but not to the leading term of the positive-signature amplitude.

With suitable choices of the Dirac γ-matrices, in sixth order the positive-signature amplitude is real and of order of $(\ln s)^2$, and the negative-signature amplitude is purely imaginary and of order of $\ln s$.

In the eighth order, seven diagrams contribute to the leading term of both the positive-signature and negative-signature amplitudes and five additional diagrams contribute to the leading term of the negative-signature amplitude. Because of these additional diagrams, there is no simple relation between the positive-signature and the negative-signature amplitudes.

In table 1, we list the number of diagrams for the 6th to 12th orders of perturbation theory. Under I, we list the number of diagrams that are important for both signatures; under II, those that contribute only to the negative-signature; and under III, those that individually contribute but the contributions cancel each other. Since 39 is an odd number, there is actually a triplet that cancel; all the others cancel in pairs.

We give the result of summing the leading terms for the negative-signature amplitude.[14] Let $\bar{M}_-$ be the Mellin transform of this amplitude

$$\bar{M}_-(\zeta) = \int_0^\infty ds\ s^{-1-\zeta} M_-(s) \ . \tag{4}$$

Then $\bar{M}_-(\zeta)$ is given by

Table 1:   Numbers of contributing diagrams to fermion
exchange at high energies.

Order	I	II	III	Total
6	3	1	0	4
8	7	5	0	12
10	15	20	6	41
12	31	72	39	142

$$\bar{M}_-(\zeta) \sim \frac{-g^2 \pi i}{\not{A}+m} \; \frac{-\zeta^{-1}\alpha(\vec{\Delta})+(\not{A}+m)\Sigma_3(\vec{\Delta},\zeta)}{1+\zeta^{-1}\alpha(\vec{\Delta})-(\not{A}+m)\Sigma_3(\vec{\Delta},\zeta)} \quad , \tag{5}$$

where

$$\alpha(\vec{\Delta}) = g^2(\not{A}+m) \int \frac{d^2\vec{k}_\perp}{(2\pi)^3} \; \frac{1}{\vec{k}_\perp^2+\lambda^2} \; \frac{\not{A}+\not{k}_\perp-m}{(\vec{\Delta}+\vec{k})^2+m^2} \quad , \tag{6}$$

$$\Sigma_3(\vec{\Delta},\zeta) = \frac{i}{\pi\zeta} \int \frac{d^2\vec{k}_\perp}{(2\pi)^3} \; \frac{1}{\vec{k}_\perp^2+\lambda^2} \; \alpha(\vec{k}_\perp+\vec{\Delta}) f(\vec{k}_\perp,\zeta) \quad , \tag{7}$$

and $f(\vec{k}_\perp,\zeta)$ is the solution of the integral equation

$$[1-\zeta^{-1}\alpha(\vec{\Delta}+\vec{k}_\perp)] f(\vec{k}_\perp,\varsigma)$$

$$= -2\pi i \; \zeta^{-1} g^2 (\not{A}+\not{k}_\perp+m)^{-1}$$

$$+\zeta^{-1}g^2 \int \frac{d^2\vec{k}_\perp}{(2\pi)^3} \frac{1}{\vec{k}_\perp'^2+\lambda^2} \; \frac{\not{A}+\not{k}_\perp+\not{k}'_\perp-m}{(\vec{\Delta}+\vec{k}_\perp+\vec{k}'_\perp)^2+m^2} \; (\not{A}+\not{k}'_\perp+m) f(\vec{k}'_\perp,\zeta).$$
$$\tag{8}$$

In the above, λ is the photon mass, m is the fermion mass, and $\vec{\Delta}$ is the momentum transfer.

The singularity in the angular-momentum space bears a remarkable similarity to that for case 1A, and is discussed in some detail in reference 14 together with some possible application.

Let us digress for a moment to relate some personal experiences in carrying out the above analysis of fermion exchange in quantum electrodynamics. Although the number of important diagrams increases rapidly with order, and the complexity of each diagram also increases with order, the amount of time actually spent on each order is roughly a constant. The reason is that, after the analysis is carried out in each order, the method of analysis can be

greatly streamlined. The total time spent is about two
years. However, the actual time of doing the work is much,
much less. Most of the time was spent in gathering courage, and in asking repeatedly the question: Shall we
do it? Shall we not do it? Shall we do it? Shall we
not do it?...

II.   NON-ABELIAN GAUGE THEORY (YANG-MILLS THEORY)

The study of the high-energy behavior, in the sense[1]
of the Yang-Mills theory[1] with Higgs mechanism was first
carried out by Nieh and Yao[15] in the sixth order. Unfortunately, they made a mistake and did not get the
correct orders of magnitude. By now a number of the
scattering processes in such a theory has been analyzed
including in particular the scattering involving an
isospin-$\frac{1}{2}$ fermion (F) and the original Yang-Mills gauge
particle described by the $\vec{b}$-field (B). We summarize
here the results up to the sixth order[16]

$$M/s \sim -2g^2 \ \frac{1}{\vec{\Delta}^2 + \lambda^2} \ T^{(1)}$$

$$+g^4 K_2 [2(\ln s - \frac{\pi i}{2})T^{(1)} + \frac{8}{3}\pi i \ T^{(0)} + \pi i \ T^{(2)}]$$

$$+g^6 \{-(\vec{\Delta}^2 + \lambda^2)[(\ln s)^2 - \pi i \ \ln s]K_2^2 T^{(1)}$$

$$-\frac{16}{3}\pi i \ \ln s [(\vec{\Delta}^2 + \frac{5}{4}\lambda^2)K_2^2 - K_3]T^{(0)}$$

$$+\pi i \ \ln s [(\vec{\Delta}^2 + 2\lambda^2)K_2^2 - 4K_3]T^{(2)}\} + \ldots, \qquad (9)$$

where λ is the mass of B due to Higgs mechanism,

$$K_2 = K_2(\vec{\Delta}) = \int \frac{d^2\vec{k}}{(2\pi)^3} \frac{1}{\vec{k}_\perp^2 + \lambda^2} \frac{1}{(\vec{k}_\perp + \vec{\Delta})^2 + \lambda^2} , \qquad (10)$$

and $K_3 = K_3(\vec{\Delta}) = \int \frac{d^2\vec{k}_\perp}{(2\pi)^3} \frac{d^2\vec{k}_\perp'}{(2\pi)^3} \frac{1}{\vec{k}_\perp^2 + \lambda^2} \frac{1}{\vec{k}_\perp'^2 + \lambda^2} \frac{1}{(\vec{k}_\perp + \vec{k}_\perp' + \vec{\Delta})^2 + \lambda^2} .$

$$\qquad (11)$$

In (9), $T^{(i)}$ means the amplitude for the exchange of isospin i; they are explicitly

$$T^{(0)} = \begin{cases} \frac{9}{256m^2} I^{(1)} I^{(2)} & \text{for FF} \\[2mm] \frac{3}{16m} \delta_{ab} & \text{for FB} \\[2mm] \delta_{ac}\delta_{bd} & \text{for BB,} \end{cases}$$

$$T^{(1)} = \begin{cases} \frac{1}{16m^2} \vec{\tau}^{(1)} \cdot \vec{\tau}^{(2)} & \text{for FF} \\[2mm] \frac{1}{8m}(\tau_a\tau_b - \tau_b\tau_a) & \text{for FB} \\[2mm] \delta_{ad}\delta_{bc} - \delta_{ab}\delta_{cd} & \text{for BB,} \end{cases}$$

and
$$T^{(2)} = \begin{cases} 0 & \text{for FF and FB} \\[2mm] \delta_{ad}\delta_{bc} + \delta_{ab}\delta_{cd} - \frac{2}{3}\delta_{ac}\delta_{bd} & \text{for BB,} \end{cases} \qquad (12)$$

where a,b,c, and d are the isospin indices, $\vec{\tau}$ are the usual Pauli matrices, and I is the identity matrix. The result for the eighth order has also been found.[17]

These results on the high-energy behavior of the Yang-Mills theory have been obtained quite recently. At present there is a great deal of activity on this subject, both here and in the U.S.S.R., and the final outcome cannot be reliably predicted. However, let us add the following speculation.

It is a feature of (9) that, for every order of perturbation, the isospin - 1 channel has one more factor of ln s than the isospin - 0 and isospin - 2 channels (in the t direction). This has very far-reaching consequences. In particular, the isospin - 1 exchange must become dominant sooner or later. The similarity between the high-energy behavior of the Yang-Mills theory and that of electrodynamics, already evident in (9), is likely to go even much further. We therefore suspect that it is more profitable to study the Abelian case in further detail than to study the more complicated non-Abelian case, at least for the limit (1) of high energies with fixed momentum transfer.

We are greatly indebted to Professor Chen Ning Yang for the most helpful discussions throughout this program.

REFERENCES

1. C. N. Yang and R. L. Mills, Phys. Rev. $\underline{96}$, 191 (1954).

2. P. W. Higgs, Phys. Lett. 12, 132 (1964) and Phys.
    Rev. Lett. $\underline{13}$, 508 (1964); F. Englert and R. Brout,
    Phys. Rev. Lett. $\underline{13}$, 321 (1964); G. S. Guralnik,
    C. R. Hagen, and T. W. B. Kibble, Phys. Rev. Lett.
    $\underline{13}$, 585 (1964).

3. T. D. Lee, Kerson Huang, and C. N. Yang, Phys. Rev.
    $\underline{106}$, 1135 (1957). See especially Sec. 4.

4. T. T. Wu, Phys. Rev. $\underline{149}$, 380 (1966). See Sec. 8(H).

5. H. Cheng and T. T. Wu, Phys. Rev. D$\underline{1}$, 467 and 2775
    (1970); G. V. Frolov, V. N. Gribov, and L. N. Lipatov,
    Phys. Letts. $\underline{31B}$, 34 (1970).

6. H. Cheng and T. T. Wu, Phys. Rev. Lett. $\underline{24}$, 1456
    (1970).

7. M. Froissart, Phys. Rev. $\underline{123}$, 1053 (1961).

8. H. Cheng, J. K. Walker, and T. T. Wu, "Impact Picture
    of Very-High-Energy Hadron Interactions (with Numer-
    ical Results)", contributed paper to the 16th Inter-
    national Conference on High-Energy Physics (1972).

9. U. Amaldi, R. Biancastelli, C. Bosio, G. Matthiae,
    J. V. Allaby, W. Bartel, G. Cocconi, A. N. Diddens,
    R. W. Dobinson, and A. M. Wetherell, Phys. Lett.
    $\underline{44B}$, 112 (1973); S. R. Amendolia, G. Bellettini,
    P. L. Braccini, G. Bradaschia, R. Castaldi, V.
    Cavasinni, C. Cerri, T. Del Prete, L. Foa, P. Giromini,
    P. Valdata, G. Finocchiaro, P. Grannis, D. Green,
    R. Mustard, and R. Thun, Phys. Lett. $\underline{44B}$, 119 (1973).

10. H. Cheng, J. K. Walker, and T. T. Wu, Phys. Lett.
    $\underline{44B}$, 97 (1973)

11. T. T. Wu and H. Cheng in "High Energy Collisions-1973"
    (ed. Chris Quigg, American Institute of Physics,
    New York, 1973), p. 54.

12. A. S. Carroll, I-H. Chiang, T. F. Kycia, K. K. Li,
    P. O. Mazur, P. Mockett, D. C. Rahm, R. Rubinstein,
    W. F. Baker, D. P. Eartly, G. Giacomelli, P. F. M.
    Koehler, K. D. Pretzl, A. A. Wehmann, R. L. Cool,
    and O. Fackler, Phys. Rev. Lett. 33, 928 and 932
    (1974).

13. M. Gell-Mann, M. L. Goldberger, F. W. Low, E. Marx,
    and F. Zachariasen, Phys. Rev. 133, B145 (1964).

14. B. M. McCoy and T. T. Wu, Phys. Rev. Lett. 35, 1190
    (1975).

15. H. T. Nieh and Y. P. Yao, Phys. Rev. Lett. 32, 1074
    (1974).

16. B. M. McCoy and T. T. Wu, Phys. Rev. Lett. 35, 604
    (1975); Phys. Rev. D12, 3257 (1975).

17. L. N. Lipatov; C. Y. Lo and H. Cheng; and B. M. McCoy
    and T. T. Wu: all unpublished.

NEW MODEL INDEPENDENT RESULTS ON THE DIFFRACTION PEAK

H. Cornille* and A. Martin

(Presented by A. Martin)

CERN - Geneva

1. INTRODUCTION

In this session we have heard very interesting talks on possible dynamical schemes which could describe high energy collisions and especially the diffraction peak, i.e., the small t large s Reggeon for elastic scattering. It may also be of some usefulness to try to obtain as much as possible rigorous information from general principles, if there is any rigorous information still to be found. It happens that during the last few months Henri Cornille and I have been able to obtain some new results, some of which, concerning the real part effects in the forward elastic peak, are already in the process of being published[1] while the study of scaling of elastic differential cross-sections, stimulated by the experimental observation of the general trend of the elastic pp diffraction peak to shrink at all values of t[2], is still unpublished. Finally, the extension of

---

*CEN - Saclay

the old inequality obtained by MacDowell and myself
to particles with spins is still, as you will see,
unfinished. We shall, however, start with the latter
item.

2.   AN INEQUALITY ON THE SLOPE
OF THE
DIFFRACTION PEAK VALID FOR ARBITRARY SPINS

In 1964, Sam MacDowell and I obtained the following
inequality on the slope of the diffraction peak, for
particles with spin 0:

$$R = \frac{36\pi \frac{d}{dt} \ell nA(s,t)\big|_{t=0}}{\frac{(\sigma_t)^2}{\sigma_{el}}} \gtrsim 1 \; , \tag{1}$$

σ_t total cross-section,
σ_{el} elastic cross-section,
$A(s,t)$ absorptive part of the scattering amplitude.
Terms of order $1/s$ have been neglected.

To obtain this inequality we used explicitly the
fact that the imaginary parts of the partial wave
amplitudes were positive. The question is now to
generalize this to particles with spins. The two-body
elastic cross-section for particles with spins is of the
form

$$\frac{d\sigma}{d\Omega} = N(s) \sum_i |F_i|^2 \; , \tag{2}$$

where the F's are amplitudes in a convenient basis
(for instance, the helicity basis or the transversity
basis) such that no cross terms are present. Thus one

can define

$$\langle A \rangle^2 = \sum_i |A_i|^2 \; , \tag{3}$$

where the A's are the absorptive parts of the various amplitudes.

We have proved from $\text{Im } T = T^+ T$ that $\langle A \rangle^2$ possesses the following positivity property

$$\langle A \rangle^2 = \sum_n c_n \cos(n\theta), \tag{4}$$

$$c_n \geqslant 0, \; \theta \text{ scattering angle.}$$

From this property <u>alone</u> we have been able to find a generalization of inequality (1):

$$R = \frac{18\pi \frac{d}{dt} \, \ell n \langle A \rangle^2 |_{t \,=\, 0}}{(\sigma_t)^2 / \sigma_{el}} > C \; . \tag{5}$$

The proof uses general properties of sums of the type (4) which hold, in fact, as shown by Henri Cornille, for a very large class of sums of orthogonal polynomials with positive coefficients[4]

At present we know with certainty that $C > 0.8$. However, a calculation in collaboration with Guy Auberson is in progress and I believe with 90% confidence that, except for terms of order $1/s$ we shall get the same answer as in the scalar case, i.e.,

$$C = 1. \tag{6}$$

To reach this goal it may be necessary to use a property

which is stronger than (4).

    Assume C = 1. When can we get R = 1? Only when
i) the imaginary part of the double flip terms vanish
in the forward direction; ii) total cross-sections are
independent of the spins on the initial particles.

    R can be estimated from experiment if one neglects
the real part of the scattering amplitude. Then we find
that R is never much bigger than unity. Typical values
are 1.2, 1.4 depending on the process. If the differentia
cross-section could be fitted by a pure exponential of
t, R would necessarily be equal to 9/8, if i) and ii)
are satisfied well enough.

3. THE QUESTION OF SCALING

3.1. The Case $\sigma_t \sim$ const. $(\log s)^2$

    This part is historical, in fact almost prehistorical
In 1970, Auberson, Kinoshita and I[5] proved the following
theorem.

    Assume that the Froissart bound is <u>qualitatively</u>
saturated, i.e.,

$$\sigma_{total} \sim \text{Const.}(\log s)^2 \ , \tag{7}$$

where the constant in front need not be the largest
possible constant. Then define

$$\tau = t(\log s)^2 \sim t\sigma_t \sim t\frac{(\sigma_t)^2}{\sigma_{el}} \ . \tag{8}$$

Then there exists a sequence of energies s_1, s_2, $s_N \to \infty$
such that, for any fixed τ,

$$\frac{\frac{d\sigma}{dt}(s_N, t)}{\frac{d\sigma}{dt}(s_N, o)} \rightarrow f(\tau) \quad . \tag{9}$$

$f(\tau)$ is an entire function of τ of order $1/2$ which has infinitely many zeros, some of which may be very close to the physical region.

All models which saturate the Froissart bound, such as the Cheng-Wu-Walker model[6] automatically possess these properties if they satisfy s channel unitarity and have good analyticity properties in t.

Perhaps you will wonder why we did this in 1970, at a time where there were no indications that cross-sections were rising. Our paper was devoted to the analysis of the violation of the Pomeranchuk theorem. However, we were wise to indicate in the last paragraph of our paper, that all the results we had obtained on the odd signature amplitude could be applied to the even signature amplitude in a situation where the cross-sections would rise like $(\log s)^2$. We know now that total cross-sections are rising. However, though they are compatible with an asymptotic $(\log s)^2$ rise they may be not rising as fast as that. A fit with $\sigma_t \sim \log s$ is perfectly compatible with the data, for instance. This is the motivation for the next subsection.

3.2. <u>The case where $\sigma_t/(\log s)^2 \rightarrow 0$</u>.

In that case we cannot establish the scaling of differential cross-sections from nothing. First one has to choose a proper scaling variable. Our choice is

$$\tau = t\frac{(\sigma_t)^2}{\sigma_{el}} \tag{10}$$

(this does not always coincide with what is called geometric scaling, where $\tau = t\sigma_t$).

In Section 2, we have defined the quantity R, and established $R \geq$ const (probably unity). Experimentall R is never much larger than unity. Then let us make the following assumptions: i) the real parts of the amplitud are negligible; ii) $R <$ const, independent of energy.

Under these assumptions we can probe again that ther is a sequence $s_1, s_2, s_N \ldots \to \infty$, such that

$$\lim_{N \to \infty} \frac{d\sigma}{dt} (s_N, t) / \frac{d\sigma}{dt} (s_N, o) \to f(\tau)$$

for any fixed real $\tau < 0$. $f(\tau)$ is a continuous, differentiable function such that $f(0) = 1$, but not identically equal to unity. $f(\tau)$ is defined only on the negative real axis. It may be impossible to continue it for complex τ.

Let us sketch the proof. Define

$$f(s, \tau) = \frac{d\sigma}{dt}(s, t) / \frac{d\sigma}{dt}(s, o) \tag{11}$$

i) for $\tau < 0$, physical, we have $|f(s, t)| \leq 1$, because of the positivity property (4):

$$\left| \sum c_n \cos n\theta \right| \leq \sum |c_n| = \sum c_n$$

ii) the derivative of a sum of the type (4) is again of the __same__ type; therefore

$$\left| \frac{d}{d\tau} f(s, \tau) \right| < \frac{d}{d\tau} f(s, o) = \frac{R}{18\pi} < \text{Const} .$$

If we take a sequence s_1, $s_2 \ldots s_N \ldots \infty$ the set of all $f(s_i, \tau)$ is a set of uniformly bounded, equicontinuous functions [such that $|f(s_i, \tau) - f(s_i, \tau')| < C|\tau - \tau'|$, C , independent of s_i]. The theorem of Ascoli-Arzela states that out of this set we can find a subset of functions approaching a limit when $s_i \to \infty$. The limit function will be called $f(\tau)$.

What can we say about $f(\tau)$? It is obviously continuous, with a bounded derivative, and it is non trivial. To prove the latter point, consider $\tau_0(s)$ such that for $\tau = -\tau_0(s)$, $f(s, -\tau_0(s)) = 1/2$; $\tau_0(s)$ defines the half width. Then we have

$$\sigma_{el} > \frac{1}{2} \frac{d\sigma}{dt}(s, t = 0) \times \frac{\sigma_{el}}{(\sigma_t)^2} \tau_0(s)$$

and using the optical theorem

$$\tau_0(s) < \text{Const} \tag{12}$$

This shows that $f(\tau)$ cannot be identically equal to unity.

Other conditions for scaling can be found. One of them is particularly appealing. Suppose we fit the small t region, for a given energy, by

$$\frac{d\sigma}{dt}(s, t) / \frac{d\sigma}{dt}(s, o) = \exp\left[b(s)t + c(s)t^2\right] ,$$

take as a scaling variable

$$\tau = b(s)t \quad ; \tag{13}$$

assume now

$$\frac{c(s)}{\left[b(s)\right]^2} < \text{Const.} \tag{14}$$

Then we get scaling in the same sense as previously.

Let me make a few final remarks on this section. We have not been able to prove the uniqueness of the limit function $f(\tau)$. There could exist another set of energies for which we get a <u>different</u> limit. Then, as we increase the energy, the amplitude would oscillate between the two limits. This could even happen if R approaches a limit as the energy goes to infinity. This incidentally is also the case when $\sigma_t \sim (\log s)^2$. Let us notice also that $\sigma_t \sim (\log s)^2$ can also be incorporated in (3.2). Indeed, it is established that

$$\frac{d}{dt}\log A(s,t)\Big|_{t=o} < (\log s)^2 .$$

Hence if $\sigma_t \sim (\log s)^2$ we also have $\sigma_{el} \sim (\log s)^2$ and R < const. follows.

4. REAL PART EFFECTS

Here we have some results which are being published[1] but these results are not yet sufficient to justify the neglect of the real part in Section 3. Let me state what we have for the scalar case.

The only control we have on the real part comes from analyticity with respect to s, i.e., dispersion relations. To use them in an efficient way we must assume that odd signature terms are negligible [this excludes in particula terms of the form s f(t) in the amplitude]. Then, if we assume, in addition

$$\sigma_t \times s^{\frac{1}{2}} \to \infty,\tag{15}$$

a very reasonable assumption, we can prove

$$(\log s)^2 > \eta = \frac{d}{dt} \log\left(\frac{d\sigma}{d\Omega}\right)_{t\,=\,o} > 0 \ ,\tag{16}$$

at least for a sequence of energies s_1, $s_2 \ldots s_N \to \infty$. This proves the existence of a forward peak of the elastic differential cross-section. If

$$\eta_{abs} = 2\,\frac{d}{dt}\,\log A(s,t)\big|_{t\,=\,0}$$

is <u>sufficiently</u> smooth $\left[\text{example } \eta_{abs} \sim C(\log s)^\gamma\right]$,

$$|\eta/\eta_{abs}| < c \quad |\eta_{abs}/\eta| < c$$

for a sequence of energies. This holds also if η (instead of η_{abs}) is sufficiently smooth.

These results are not sufficient to treat the problem of scaling with real parts. There we would have to show, for instance, in the case of $\sigma_t \sim (\log s)^\gamma$, that there is a sequence of energies for which, for <u>all</u> $-T \le t \le 0$, the real part is small, compared to the imaginary part. What is relatively easy to do is to show that for one given value of t there is a sequence of s_i's for which the real part is small. If we change t the new sequence may have no common point with the previous one. This is the problem.

The assumption that the odd signature amplitude is neglible is unavoidable. Otherwise it is easy to build counter-examples with a dip instead of a peak in the forward direction. I do not believe that odd signature amplitudes are important at high energies; I even made a bet of a *bottle of champagne* about this with Elliott Leader. However, I must frankly admit that there is no fundamental objection against the presence of such terms.

REFERENCES

1. H. Cornille and A. Martin, CERN preprint TH. 2045
    (1975), to appear in Nuclear Physics B.

2. See, for instance, the proceedings of the Palermo
    Conference (1975).

3. S. W. MacDowell and A. Martin, Phys. Rev. 135B, 960
    (1964).

4. H. Cornille, in preparation.

5. G. Auberson, T. Kinoshita and A. Martin, Phys.
    Rev. D3, 3185 (1971).

6. H. Cheng, J. D. Walker and T. T. Wu, Phys. Letters
    44B, 97 (1973); Phys. Letters 44B, 283 (1973).

Some of the participants of the Orbis Scientiae from the Soviet Union, Professors Vladimirov, Slavnov, Bogolubov and Matveev.

S-CHANNEL UNITARITY IN THE POMERON THEORY WITH $\alpha_P(0) > 1$.

M. S. Dubovikov, K. A. Ter-Martirosyan

(Presented by K. A. Ter-Martirosyan)

Institute for Theoretical and Experimental Physics

Moscow, U. S. S. R.

ABSTRACT

The integral equation defining the scattering amplitude in the form of the sum of all pomeron graphs contributions is obtained in the theory with $\alpha_P(0) > 1$. It is shown that for definite restrictions on the Cardy's vertex $G_{oo}(k^2)$ of the pomeron coupling, the solution of this equation in the ξ, b - representation $\xi = \ln(s/m)^2$ is the simple amplitude $\Theta(a\xi - b)$, with $a = 4\alpha_P'.\Delta$, $\Delta = \alpha_P(0)-1$. It leads to the Froissart type rising cross sections, and diffraction cone slopes: $\sigma^{tot} \sim B \sim \xi^2$.

The theory in its simple version at $G_{oo}(k^2) = 0$, yields a good description of existing experimental data at $\Delta = 0.07-0.08$, at $\alpha_P' \simeq 0.3$ and at the values of pomeron to particle coupling vertices given in the table II of the paper. It leads to the prediction of very large values of $\sigma^{tot}_{NN} \simeq \sigma^{tot}_{kN}$ $\sigma^{tot}_{\pi N} \simeq 100-150$ mb in the range of energies $E_{LAB} \sim 10^5-10^{10}$ GeV available in the near future.

It is shown below that the pomeron theory with $\alpha_P(0) > 1$[1] can satisfy s-unitarity condition under definite restrictions on the pomerons coupling vertex $G_{oo}(k^2)$, introduced by Cardy[1]. To this end we propose the method of summation of all Gribov-Cardy pomeron graphs[2] and show that the solution of the corresponding integral equation is, as $s \to \infty$, the simple amplitude $\Theta(a\xi - b)$ for the diffraction on the sphere with sharp edge. In accordance with Abramovsky, Gribov and Kancheli's results[3] the s-channel unitarity is supposed to be the equivalent to the existence of the summed contributions of all "enhanced"[2] pomeron graphs. For the sake of the simplicity we shall use below the eikonal approximation, although our results are of a general nature and can be extended to the wide class of the theories[1] considered by Cardy.

The scattering amplitude is represented by the graphs shown in fig. 1. In ξ, b - representation, where $\xi = \ln (s/2m_N^2)$, it corresponds, in the eikonal model, to the amplitude which we shall call the froissaron:

$$F(\xi, b) = 1 - e^{-v(\xi, b)},\qquad (1)$$

where $v(\xi, b)$ is given by the contribution of all graphs irreducible against the "eikonalization" (that is those which do not form of one of the terms in the eikonal set of graphs, except for the first one), $v(\xi, b)$ is the sum

$$v(\xi, b) = v_0(\xi, b) + C(\xi, b) + D(\xi, b),\qquad (2)$$

where

$$v_0(\xi, b) = \frac{G\overset{o}{a}\; G\overset{o}{b}}{\alpha_P'\, \xi}\, e^{\xi\Delta\, -\, b^2/4\alpha_P'\, \xi}\qquad (3)$$

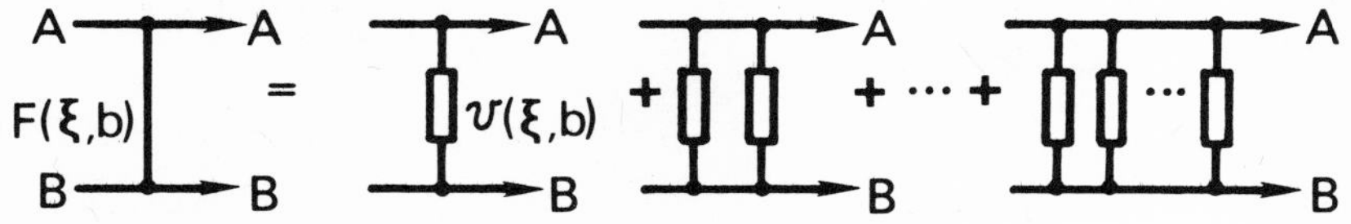

Figure 1

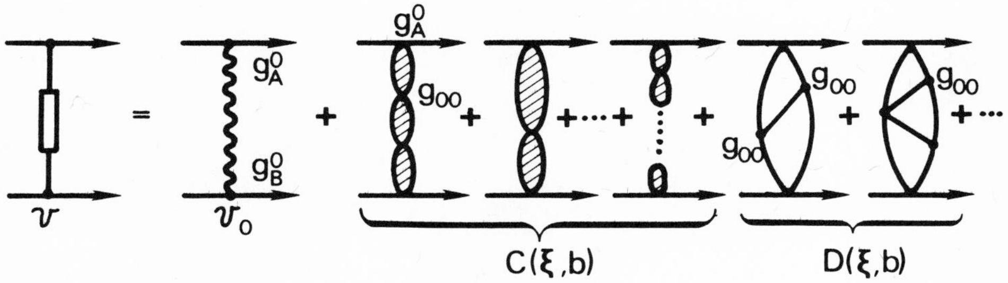

Figure 2

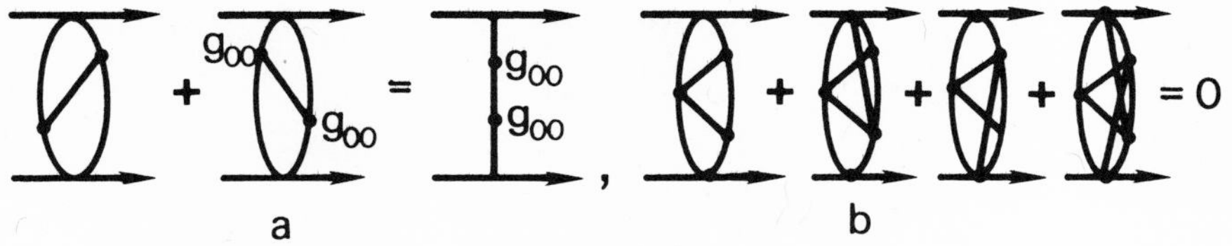

Figure 3

is the Green's function of the pomeron with $\alpha_p(0)-1=\Delta>0$,
$C(\xi, b)$ is the contribution of all irreducible graphs
of the form of the chain along the t-channel, with the
links connected only by the vertices $G_{oo}(k^2)$ (and with
the number of them larger than one). $D(\xi, b)$ corresponds
to all irreducible graphs not having the form of a chain
and does not include the pomeron contribution $v_o(\xi, b)$.
Eq. (2) is represented graphically by fig. 2.

Obviously, each term in the set $D(\xi, b)$ has a form
of an irreducible graph with more than three lines entering each vertex with the froissaron Green's function
(1) corresponding to each line, and the coupling constant
$G_{oo}(k^2)$ corresponding to each vertex. The simplest graphs
from the set $D(\xi, b)$ are shown on fig. 2.

In $\omega, \vec{k}$-representation (where $\omega = j - 1$, $\vec{k} = q\vec{\perp}$
$t \simeq - q_{\perp}^2$) the chain contribution is the sum of the geo-
metrical progression

$$C(\omega, k^2) = \frac{G_{oo}(k^2)\, Z^2(\omega, k^2)}{1 - G_{oo}(k^2)\, Z(\omega, k^2)} , \qquad (4)$$

where $Z(\omega, k^2)$ corresponds to the link of the chain. The
latter is represented by the set of the same graphs as
the whole froissaron, except for the chain, i.e.

$$Z(\omega, k^2) = F(\omega, k^2) - C(\omega, k^2). \qquad (5)$$

Inclusion of the chain $C(\omega, k^2)$ in $Z(\omega, k^2)$ would lead
to overcounting the number of graphs.

Inserting (5) into (4) we get the algebraic equation
for $C(\omega, k^2)$. Its solution is

$$C(\omega, k^2) = \frac{G_{oo}(k^2)\, F^2(\omega, k^2)}{1 + G_{oo}(k^2)\, F(\omega, k^2).} \qquad (6)$$

Eqs. (1), (2), (6) can be considered as the integral equation for the function $F(\xi, b)$, as $C(\xi, b)$ is
expressed through $F(\xi, b)$ in ω, k-representation by
eq. (6) and the term $D(\xi, b) \to D(\omega, k^2)$ is connected to
$F(\omega, k^2)$ through the Gribov-Cardy graphs calculus [1,2]
rules.

Let us show that this equation has as $\xi \to \infty$ the
sharp edge solution

$$F(\xi, b) \simeq \Theta(a\xi - b), \qquad\qquad (7)$$

which corresponds to a scattering amplitude of the form
$T(s,t)/s \simeq ib_o J_1(kb_o)/k$, with $b_o \simeq a \cdot \xi$.

Leaving in (2) only the first term $v \simeq v_o$ (which
corresponds to the approximation considered [1] by Cardy)
we obtain at $\xi \to \infty$ the solution (7), with

$$a^2 = 4\alpha'_P \cdot \Delta. \qquad\qquad (8)$$

In fact, at $C = D = 0$ eq. (1) gives the "smearing"
Θ -function, differing from (7) in the region $b \simeq a\xi$
at the width $\delta b \simeq \sqrt{\alpha'/\Delta}$ (as here $v_o(\xi, b) \simeq \dfrac{G^o_a G^o_b}{\alpha'_P \xi} \exp\left[\dfrac{a\xi - b}{\sqrt{\alpha'/\Delta}}\right]$

However, the scattering amplitude $T(s,t)/s$ remains
practically the same for $b_o = a \cdot \xi \pm \delta$ (as for $b_o = a\xi$)
in the essential region inside the diffraction cone
$k \lesssim 1/a \cdot \xi$

The solution (7), (8) is reproduced by the right
hand part of the eq. (1), (2) and (6), only if terms
C, D in (2) i) tends to zero at $b > a \cdot \xi$, and ii) are
either both positive, or have modulae less than at
$b < a \cdot \xi$ (and as $\xi \to \infty$).

If (7) turns out to be the solution, then only
graphs from fig. 3a contribute to $D(\xi, b)$.   The

contribution of graphs with more than two vertices
vanishes [1]; for example the contributions of graphs
shown in fig. 3b cancel each other. Note, moreover,
the equality of graph contributions shown in fig. 3a.
Consequently, the function $D(\xi, b)$ is both positive
and finite at $b < a\xi$ and equal to zero at $b > a\xi$.

Thus we have to consider the behavior of the chain
contribution $C(\xi, b)$. It is obvious from (6) that
$C(\xi, b) = 0$ at $b > a\xi$ for the Θ -function-like solution
(7). To analyze $C(\xi, b)$ in the region $b < a\xi$ let
us insert in (6) the function

$$F(\omega, k^2) = a^2(\omega^2 + a^2 k^2)^{-3/2}, \tag{9}$$

which is the exact ω, k transform of the Greens function
(7). It gives

$$C(\omega, k^2) = \frac{a^2}{(\omega^2 + a^2 k^2)^{3/2}} - \frac{a^2}{(\omega^2 + a^2 k^2)^{3/2} + G_{oo}}. \tag{10}$$

The contribution of the first term to $C(\xi, b)$ is
positive and not essential. The second term has in the
right - half plane the Regge poles

$$\omega_{1,2} = \sqrt{\eta_{1,2}^2 \; G_{oo}^{2/3}(k^2) - a^2 k^2}, \qquad \eta_{1.2} = e^{\pm\frac{2\pi i}{3}} \tag{11}$$

for $G_{oo} > 0$, or

$$\omega_1 = \sqrt{|G_{oo}(k^2)|^{2/3} a^2 k^2} \tag{12}$$

for $G_{oo} < 0$.

As $\xi \to \infty$ these poles give the main contribution

$$C(\xi,k^2)=-\frac{2}{3}\,Re\,\frac{a^2\exp[(G_{oo}^{2/3}\eta_1^2-a^2k^2)^{1/2}\xi]}{G_{oo}^{1/3}\eta_1(G_{oo}^{2/3}\eta_1^2-a^2k^2)^{1/2}} \quad (13)$$

for $G_{oo}>0$, or

$$C(\xi,k^2)=-\frac{1}{3}\,\frac{a^2\exp[(|G_{oo}|^{2/3}-a^2k^2)^{1/2}\xi]}{|G_{oo}|^{1/3}(|G_{oo}|^{2/3}-a^2k^2)^{1/2}} \quad (14)$$

for $G_{oo}<0$.

The transition into the b-representation requires the computation of contour integrals in the complex k-plane by the saddle point method. For the case of constant G_{oo} not depending on k^2 it gives:

$$C(\xi,b)=-\frac{2}{3}Re\,\frac{\Theta(a\xi-b)\exp[G_{oo}^{1/3}\eta_1-(\xi^2-\frac{b^2}{a^2})^{1/2}]}{G_{oo}^{1/3}\eta_1(\xi^2-b^2/a^2)^{1/2}} \quad (15)$$

for $G_{oo}>0$, or

$$C(\xi,b)=-\frac{1}{3}\,\frac{\Theta(a\xi-b)\exp[|G_{oo}|^{1/3}(\xi^2-\frac{b^2}{a^2})^{1/2}]}{|G_{oo}|^{1/3}(\xi^2-b^2/a^2)^{1/2}} \quad (16)$$

for $G_{oo}<0$.

At fixed $\xi-b/a=$ Const and at $\xi\to\infty$ the value of $|C|$ increases like $e^{A\sqrt{\xi}}_\xi$, becoming larger than $v_o(\xi,b)$ at sufficiently large ξ and leading, due to (1), to the violation of s-unitarity, however, the allowance of the weak dependence of G_{oo} on k^2 of the type $G_{oo}=\pm G_o-G_1k^2$, with G_o, $G_1>0$ changes the situation. Eqs. (13), (14)

show that it leads to decreasing the value of the constant a in (15), (16): $a^2 \to a_1^2 = a^2 - C_o g_1$, where C_o is
some constant. Thus, the exponent in (15), (16) will
now vanish, together with $C(\xi, b)$, at $b = a_1 \xi < a\xi$ giving
$|C(\xi, b)| < v_o(\xi, b)$ for all $b < a\xi$ at the following upper
bounds for $G_o = |G_{oo}(0)|$:

$$G_o^{1/3} < 2\Delta \quad \text{for} \quad G_{oo} > 0,$$

$$G_o^{1/3} < \Delta \quad \text{for} \quad G_{oo} < 0.$$

This means that (7) is the solution of eqs. (1),
(2) and (6) as $\xi \to \infty$, satisfying the s-channel unitarity.

Let us note that amplitude (1)-(3) near
zero, $G_{oo} \simeq 0$ gives a good description [4,5] of existing high energy data, in particular the rising cross
sections σ^{tot} and diffraction cone poles $B = B(\xi)$ (of
course, with the obvious account of the nonpomeron singularities $a \neq P$).

With the pomeron-particle coupling vertices choosen
in the simplest form $G_A(k^2) = G_A^o \exp(-R_A^2 k^2)$, eqs. (1)
- (3) yields for the pomeron part σ_P of σ^{tot} and for
$B(\xi)$ $\left(\text{defined so that } \frac{d\sigma}{dt} = \left(\frac{d\sigma}{dt}\right)_{t=o} e^{Bt}\right)$:

$$\sigma_P(\xi) \simeq \text{Im} \frac{T(s,0)}{s} = 8\pi G_A^o G_B^o \, f(z) e^{\xi\Delta},$$

$$B(\xi) = 2\text{Re} \left[\frac{d(T/s)}{(T/s)d\xi}\right]_{k^2=0} = 2(\alpha_P^{'}\xi + R^2)\tilde{f}(z)/f(z),$$

(18)

where $f(z) = \sum_n \frac{(-z)^{n-1}}{n \cdot n!} \simeq \frac{\ln(C_o z)}{z}$, $\tilde{f}(z) = \sum_n \frac{(-z)^{n-1}}{n^2 n!} \simeq \ln(C_o z)$

are decreasing functions of the variable*

$$z = \frac{G_A^\circ G_B^\circ \exp(\xi\Delta)}{\alpha_P' \cdot (\xi - \frac{i\pi}{2}) + R^2}, \quad R^2 = R_A^2 + R_B^2 .$$

Eqs. (18) contain four parmeters: $\Delta = \alpha_P(0) - 1, \alpha_P'(0)$, the value $\gamma_{AB} = G_A^\circ G_B^\circ$ and the radius R^2 of the pomeron residue.

Let us show that they can be obtained directly from the experimental data for $\sigma^{tot} \simeq \sigma_P$ and $B(\xi)$. The latter give, in the region $E_{LAB} \sim 10^2 - 10^3$ GeV (i.e. at $\xi \approx 6 \pm 1$),

$$\frac{\sigma^{tot\,\prime}(\xi)}{\sigma^{tot}(\xi)} \simeq \frac{B'(\xi)}{B(\xi)} \simeq 0.07 - 0.08.$$

According to (18) this will take place at a constant value of z, i.e. at

$$\frac{z'}{z} \simeq \Delta - \frac{\alpha_P'}{\alpha_P' \xi + R^2} \simeq 0 \tag{19}$$

at $\xi \simeq 6$. This yields: $\dfrac{\sigma_P'(\xi)}{\sigma_P(\xi)} \simeq \dfrac{B'(\xi)}{B} = \Delta$, and therefore

$$\Delta = 0.075 \pm 0.005.$$

---

*These expressions correspond to the well-known eikonal form of the scattering amplitude (1)-(3)

$$T(s,t)/8\pi s = G_A^\circ G_B^\circ e^{\xi\Delta} \sum_n \frac{(-z)^{n-1}}{n \cdot n!} \exp[-\lambda k^2/n],$$

where $\lambda = \alpha_P' \cdot (\xi - \frac{i\pi}{2}) + R^2$, $k^2 \approx -t$.

T a b l e I

Pole parameters

a	P	f	ω	ρ	A_2
$\alpha_a(0)$	1.07	0.45	0.425	0.49	0.35
$\alpha_a'(0)$ $(GeV/c)^{-2}$	0.254	0.7	1.0	0.7	0.7

T a b l e II

Parameters of the vertices

	N		K		π	
	g $(GeV/c)^{-1}$	R^2 $(GeV/c)^{-2}$	g $(GeV/c)^{-1}$	R^2 $(GeV/c)^{-2}$	g $(GeV/c)^{-1}$	R^2 $(GeV/c)^{-2}$
P	1.89 1.92	1.79 1.78	1.03 1.01	0.19 0.26	1.14 1.12	0.57 0.64
f	3.01 3.10	1.52 1.41	0.71 0.69	-1.72 -1.7	1.37 1.34	-0.18 -0.13
ω	1.83 1.86	3.86 3.76	0.59 0.58	1.99 2.08	-	-
ρ	0.30 0.30	2.0	0.75	2.0	1.51	4.56
A_2	0.45	1.0	0.60	1.0	-	-

In the essential region of $z \sim 1.2 - 1.5$, (see below), the function $\beta(z) = \tilde{f}(z)/f(z)$ has a value close to 1.2 and the quantity $\lambda = \alpha_P' \xi + R^2 \simeq B(\xi)/2.4$ at $\xi \simeq 6$ is close to 5 (as here $B(\xi) \simeq 12(\text{GeV/c}^{-2})$). Therefore (20) gives

$$\alpha_P' \simeq (6\alpha_P' + R^2) \cdot \Delta \simeq 4.5 \cdot 0.075 \simeq 0.3$$

and equation $6 \cdot \alpha_P' + R^2 \simeq 4.5$ yields $R^2 \simeq 3.2$. The parameter $\gamma_{AB} = G_A^o G_B^o$ can be found from eq. (18) and experimental data for $\sigma^{tot}(\xi)$. For the case of pp interaction it gives $\gamma_{pp} \simeq 4$ (all values are given in $(\text{GeV/c})^{-2}$; note that $8\pi(\text{GeV/c})^{-2} \simeq 9.8$ mb and $e^{\epsilon\Delta} f(z) \simeq 1$).

More accurate values of these and other parameters are given in Tables I. II. They are defined [4] from the condition of the best fit to the experiment (e.g. from minimum of χ^2) with account of all aP^n singularities, where $a = f, \omega, \rho, A_2$ (see also [5]). The corresponding σ^{tot} rises quickly (see fig. 4) having relatively large, universal values-the same for all particles - $\sim 100 - 120$ mb at $E_{LAB} \sim 10^7 - 10^{10}$ GeV. The diffraction cone slopes are not given here; they are also in good correspondence with the experimental data, and rise at $\xi > 1/\Delta$ as ξ^2.

The description of the differential cross section behavior $d\sigma/dt = \psi(t)$ in the region of not very small $t \sim 1 - 3 (\text{GeV/c})^2$ turns out to be a more difficult problem. It probably cannot be solved with the simplest exponential form of the vertices $G_A \simeq G_A^o \exp(-R_A^2 k^2)$.

It is interesting to understand how these results will change by taking into account in (1)-(3), the terms

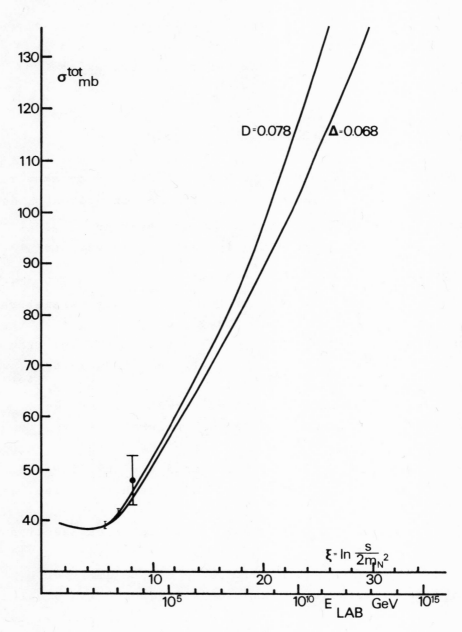

Figure 4

of the first order in the pomeron to pomeron coupling
constant $G_{oo}(k^2)$.

Consider particle production processes at $\alpha_p(0)>1$.
Their cross sections are given by different terms in
$\text{Im} T(s,o)s$, where $T(s,t)$ is represented, in the approximation $G_{oo}\simeq0$, by the multipomeron exchange[1]
graphs (fig. 5). The AGK [3] cutting rules then yield
for the central part, $y=y_{cm}=y_{lab}-\xi/2\sim1$, of the inclusive
spectrum (fig. 6)

$$\frac{d\sigma}{dy} \simeq \sigma_o u_o e^{\xi\Delta}, \quad \sigma_o=8\pi G_A^{\circ}G_B^{\circ},$$

at $\xi>>1/\Delta$. The same rules give the cross sections S_n
of the simultaneous production of n multiperipheral
showers (fig. 7), in the eikonal approximation, by [6]

$$S_n = \frac{\sigma_o e^{\xi\Delta}}{2zn} e^{-2z} \sum_{k>n} \frac{(2z)^k}{k!},$$

with the value of $z\sim e^{\xi\Delta}$ as used above. This yields for
the average number $\langle n\rangle$ of showers the value

$$\langle n\rangle = \frac{1}{\sigma_{in}} \sum_{n=1}^{\infty} n S_n = \frac{\sigma_o e^{\xi\Delta}}{\sigma_{in}} \simeq \frac{1}{f(z)} \sim \frac{e^{\xi\Delta}}{\xi^2},$$

increasing, as $\xi\to\infty$, as $\sigma_{in}=\sum_{n\geq1} S_n = \sigma^{tot}-c\sigma_{in}\simeq\sigma_o e^{\xi\Delta}f(z)\sim\xi^2$.
Account of energy conservation shows that each shower in
fig. 6 takes the energy $\sim \sqrt{s}/n$ and contributes to $\frac{d\sigma}{dy}$
only at $|y|<\ell n(\sqrt{s}/n)$, i.e. at $|y|<\xi/2-\xi\Delta$ for the essen-
tial values of $n\simeq\langle n\rangle$ (as $\ln \langle n\rangle \simeq \xi\Delta$ for $\xi\to\infty$). At the
edges $|y|>\xi/2-\xi\Delta$ the spectrum of produced particles

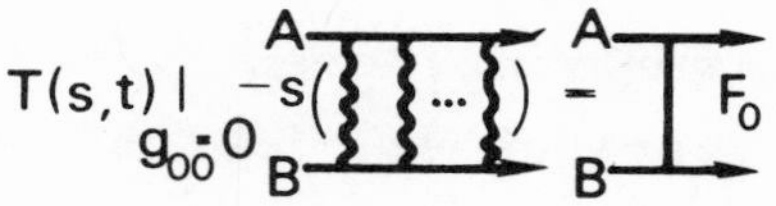

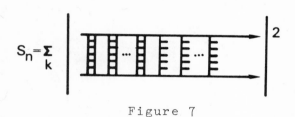

Figure 5 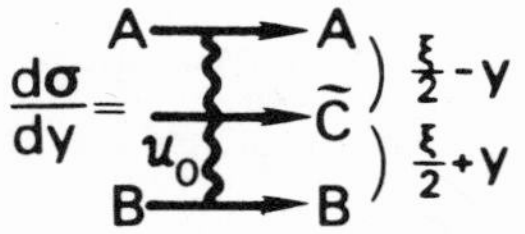   Figure 6

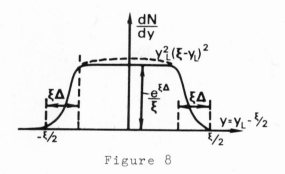

Figure 7

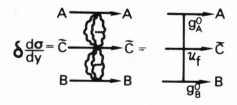

Figure 8

Figure 9

$\frac{dN}{dy} = \frac{d\sigma}{\sigma_{tot}dy}$ fall off rapidly to $\frac{dN}{dy} \to$ const at $|y| \to \xi/2$.

It has the shape shown in fig. 8 and satisfies the

the energy conservation condition $2\int_0^{\xi/2} \frac{dN}{dy} e^y dy \simeq e^{\xi/2}$ (as

the plateau region $|y| < \xi/2 - \xi\Delta$ always gives small con-
tribution to this integral at $\xi \to \infty$).

The average multiplicity of produced particles,

$$\bar{N} \simeq <n> (a\xi+b) = \frac{1}{f(z)}(a\xi+b) \to \frac{N_o e^{\xi\Delta}}{\xi + R^2/\acute{\alpha}_P} \quad ,$$

will then rise as $s \to \infty$ as a power of the energy; here

$N_o = \frac{G_A^o G_B^o}{\acute{\alpha}_P \Delta}(a+b/\xi)$ (in the eikonal approximation). Existing

cosmic ray experimental data for $E_{LAB} \sim 10^4 - 10^8$ GeV give
some preliminarary evidence for this small rise of $\bar{N}$
above the straight line $N_o(\xi) = (a\xi+b)/f(z_o)$.

At $G_{oo} \neq 0$ a small correction to $\frac{d\sigma}{dy}$ comes from two
froissaron graphs fig. 9

$$\frac{d\sigma}{dy} \simeq \sigma_o [u_o e^{\xi\Delta} + u_f y_L^2 (\xi-y)^2], y_L = y+\xi/2,$$

leading to a small curvature of the central part of the
spectrum (dotted line on fig. 8). Here u_f is the
constant of the "vertex" radiation [3]. At $\xi > 1/\Delta$ these
corrections are always negligible and, in general,
$u_o = Const \neq 0$.

Authors express their thanks to K. G. Boreskov,
V. N. Gribov, P. E. Volkovitsky, B. Z. Kopeliovicn,
A. M. Lapidus, L. I. Lapidus, V. I. Lisin and to A. A.
Migdal for the series of interesting discussions very
important for this work.

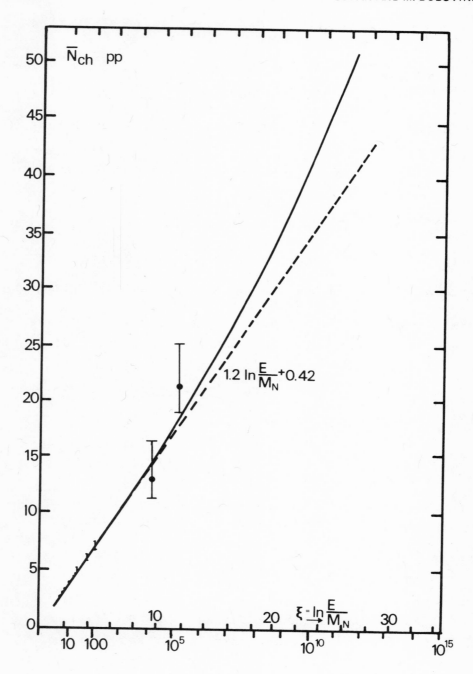

Figure 10

REFERENCES

1. J. L. Cardy, Nucl. Phys. B75, 413 (1974).
2. V. N. Gribov, JETP 53, 654 (1967).
3. V. A. Abramovski, V. N. Gribov, O. V. Kancheli Proc. of XVI Intern. High Energy Conf. Chicago-Batavia V IV (1972); Yad. Phiz. 18, 595 (1975).
4. P. E. Volkovitsky, A. M. Lapidus, V. I. Lisin, K. A. Ter-Martirosyan, Yad. Fiz. 14, 814 (1971).
5. A. Capella, J. Tran Thanh Wan, J. Kaplan, Preprint LPTHE 75/12.
6. K. A. Ter-Martirosyan, Phys. Lett., 44B, 377 (1973) (see also D. R. Snider, H. W. Wyld, Jr. Phys. Rev. D11, 2538 (1975)).

EQUATION FOR DENSITY OPERATOR AND SCALING PROPERTIES
OF INCLUSIVE CROSS SECTION

A. A. Arkhipov, A. A. Logunov, V. I. Savrin

(Presented by V. I. Savrin)

Institute for High Energy Physics

Serpukhov, U.S.S.R.

ABSTRACT

An equation for density operator of inclusive reaction that is connected with inclusive cross section via
a simple relation has been derived in Sections II - V of
the paper in the framework of equal-time formulation of
the many-body problem in quantum field theory. The
solution of this equation allows one to express the inclusive cross section through an effective quasipotential
of screening. The results obtained are similar to
scattering on a composite nucleon and is very close in
spirit to the parton idea.

In Sections VI and VII we managed to reveal a scaling character of inclusive cross section behavior as
well as a power law decrease with increase in transverse
momentum, with the help of some model assumptions with
respect to the screening quasipotential.

I. INTRODUCTION

The problem of describing a system of many

relativistic particles is one of the most important problems in Quantum Field Theory. As is known, Quantum
Field Theory is notable for the functional character of
its equations. Therefore, any problem, even the two-
particle one, turns out to be in this case a problem with
an infinite number of degrees of freedom. With contemporary mathematical techniques one cannot solve these
functional equations for the general case. For this
reason in quantum theories with strong coupling where
perturbation theory cannot be applied, it is expedient to
construct simpler dynamic equations that may serve as a
basis for model constructions and which would allow one
to describe a system of interacting particles.

In particular, a wave function for a system of two
interacting particles satisfies a quasipotential [1] that
is a relativistic analog of the Schrödinger equation.
The quasipotential serves as a kernel of this equation.
It is a complex function dependent on particle energy
and is a generalization of an ordinary potential in
nonrelativistic quantum mechanics. The imaginary part
of the quasipotential includes a contribution from all
the multiparticle intermediate states and should be
positive due to the unitarity condition. The quasipotential approach becomes simpler and illustrative in the
framework of equal-time formulation of the two-body
problem in Quantum Field Theory.[2] This is the case when
the analogy with nonrelativistic mechanics becomes
evident. Of course a quasipotential for the interaction
of two relativistic particles remains unknown; however,
analogy with quantum mechanics allows one to make a
number of model assumptions so as to construct this
quasipotential and describe experimental data.[3]

When studying multiparticle production processes we
come to the many-body problem in Quantum Field Theory,
therefore, description of such processes becomes more
complicated. The thing is that as compared with elastic
processes the number of variables in this case increases
greatly, and construction of any scheme to calculate the
corresponding matrix elements becomes considerably complicated. Nevertheless it seems possible to generalize
the quasipotential approach for the case with three [4], [5]
or even more interacting particles by constructing a
relativistic analogue of the Schrödinger equation for a
wave function of a system with a fixed number of particles.
Here a multiparticle quasipotential is not a sum of quasipotentials for interaction between separate pairs of
particles. Besides, description of inelastic processes
at high energy is complicated by the fact that the number of possible channels of the reaction is extremely
large and it is very difficult to select processes with
a fixed number of particles in the final state. In this
case one is interested in momentum distribution of, let
us say, one or two particles in the final state and having
no information on the remaining secondary hadrons (inclusive reactions). To study these reactions one has to
introduce some density operator that would describe behavior of the selected particles in the selfconsistent
field of all the other remaining particles in the final
state. Then it is possible to reduce the multiparticle
problem to the one of scattering of a small number of
particles in some fictitious field.

A formal connection of the density operator with
inclusive and exclusive cross sections was considered in
ref.[6] In our paper we shall use the quasipotential wave
function within the framework of equal-time formalism to

construct a density operator. The density operator
constructed with such wave functions is first of all
connected with the inclusive cross section through a
simple relation, and then it satisfies quasipotential
equation that is a relativistic analog of the one for a
nonrelativistic density operator describing the behavior
of a chosen particle in the selfconsistent field of the
other remaining particles in the system.

Certainly the effective quasipotential for interaction between the selected particle and selfconsistent
field remains unknown, and again an analogy with quantum
mechanics allows us to use model representations to construct this quasipotential as well as to apply usual
quantum mechanical methods for studying the selfconsistent
field. Still, even the most general form of the results
obtained in this way leads us to the picture of interaction very close in spirit to the parton idea.

In Sections VI and VII of the paper we show as an
illustration of the way in which some model assumptions
on the screening quasipotential lead to the scaling
behavior of single-particle inclusive cross section as
well as to the power-law form of its decrease at large
transverse momentum.

II. EQUATION FOR DENSITY OPERATOR OF PURE
TWO-PARTICLE STATE

A system of two interacting particles with momentum
$\vec{c}$ and energy E is described with the wave function
$\Psi_{\vec{c}}(\vec{k};t)$ that satisfies as usual quasipotential equation:

$$(E_k - E)\Psi_{\vec{c}}(t) = V(\vec{c})\Psi_{\vec{c}}(t),\qquad\qquad (1)$$

where $E_k = \sqrt{k^2+\mu^2} + \sqrt{(\vec{c}-\vec{k})^2+m^2}$, $V(\vec{k};\vec{k}'|\vec{c})$ is quasipotential

for interaction between particles with masses μ and m.
The density operator of a pure two-particle state may
be constructed according to usual rules of quantum
mechanics in the form of a product:

$$\rho(\vec{k};\vec{\kappa}|t) = \Psi_{\vec{c}}(\vec{k};t)\overset{*}{\Psi}_{\vec{c}}(\vec{\kappa};t). \tag{2}$$

As the dependence of the wave function upon time is
trivial in this case,

$$\Psi_{\vec{c}}(\vec{k};t) = \Psi_{\vec{c}}(\vec{k};o)e^{-iEt}, \tag{3}$$

we will take the wave functions in the definition of the
density operator (2) at different energies $E{\neq}E'$ such
that the operator $\rho(t)$ will depend on time in the following way:

$$\rho(\vec{k};\vec{\kappa}|t) = \rho(\vec{k};\vec{\kappa}|o)e^{-i(E-E')t}. \tag{4}$$

As can easily be shown due to Eq. (1) density operator
(2) satisfies the equation,

$$(E_k-E)\rho(t)-\rho(t)(E_\kappa-E') = V\rho(t)-\rho(t)\overset{+}{V}, \tag{5}$$

that is, an analog to the corresponding equation in nonrelativistic quantum mechanics. In the present paper we
shall show that density operator of the two-particle system with the given momentum $\vec{c}$ selected from a system with
an arbitrary number of interacting particles, produced
in two-body collision satisfies an equation similar to
Eq. (5), but with a distorted screened interaction quasipotential.

III. EQUATION FOR DENSITY OPERATOR
OF A PURE MULTIPARTICLE STATE

For definitness we shall consider the following process. Let us have meson and scalar nucleon with momenta $\vec{q}_1$ and $\vec{q}_2$, respectively, in the initial state, and in the final state n identical mesons with momenta $\vec{k}_i (i = 1,\ldots,n)$ and one nucleon with momentum $\vec{p}_2$. By analogy with two- and three-particle cases considered in refs. /2,4,5/ we shall present the wave function for the multiparticle system produced in the collision as

$$\Psi_{\vec{Q}}(y,x_1,\ldots x_n) = <o|T\left(\psi(y)\psi(x_1)\ldots\psi(x_n)\right)|\vec{q}_1\vec{q}_2>, \qquad (6)$$

where $\psi(y)$ and $\Psi(x_i)$ are operators of nucleon and meson fields in Heisenberg representation and $|\vec{q}_1,\vec{q}_2>$ is the initial state vector in the same representation, $\vec{Q} = \vec{q}_1+\vec{q}_2$ is total momentum of the system.

We shall present the equal-time wave function in the momentum space in the following way:

$$(2\pi)^3 \, \delta(\vec{p}_2+\sum_{i=1}^{n}\vec{k}_i-\vec{Q})\Psi_{\vec{Q}}(\vec{k}_1,\ldots\vec{k}_n;t) = \qquad (7)$$

$$=\sqrt{\frac{2p_2^{\,o}2k_1^{\,o}\ldots2k_n^{\,o}\,2q_1^{\,o}2q_2^{\,o}}{n!\,(2\pi)^{3(n-1)}}}<o|T\left(\psi(\vec{p}_2,t)\psi(\vec{k}_1,t)\ldots\psi(\vec{k}_n,t)\right)|\vec{q}_1\vec{q}_2>,$$

where $\vec{p}_2=\vec{Q} - \sum_{i=1}^{n}\vec{k}_i$. The wave function introduced here is remarkable for defining asymptotically the amplitude of the described process of multiparticle production:

$$\lim_{t\to\infty}\Psi_{\vec{Q}}(\vec{k}_1\ldots\vec{k}_n,t)e^{iE_kt}= 2\pi i\delta(E_k-E)\sqrt{2q_1^{\,o}2q_2^{\,o}}\, T_n(\vec{k}_1\ldots\vec{k}_n,\vec{q}_1|\vec{Q}),$$
$$(8)$$

where $E_k=p_2^{\,o} + \sum_{i=1}^{n} k_i^{\,o}$, $E=q_1^{\,o} + q_2^{\,o}$, T_n is the production amplitude for n mesons in a meson-nucleon collision.

In refs.$^{/2,5/}$ problems of elastic scattering of two
and three relativistic particles as well as elastic
scattering of a particle on the composite system were
considered. With the help of the basic property of the
free wave function and for the total wave function in
coordinate space, we have managed to obtain the equation
containing only one common time for all the particles in
the system. This equal-time quasipotential equation is
a relativistic analog to the Schrödinger equation. In
the present paper we shall perform a generalization of
the equal-time formalism for the case of a multiparticle
system, and with the help of equal-time reduction to
obtain an equal-time equation for the wave function of
the system with arbitrary fixed number of interacting
particles.

However we are not going to present here all the
calculations, we shall just formulate the basic statement
that the positive-frequency projection for the wave function (7), $\Psi^1_{\vec{Q}}(t)$ of n+1 interacting particles, introduced
in the present paper, satisfies the equal-time equation:

$$(E_{,k}-E)\Psi'_{\vec{Q}}(t) = V_n\Psi'_{\vec{Q}}(t), \qquad (9)$$

where V_n is the quasipotential for interaction of n+1
hadrons and the positive-frequency projection $\Psi'_{\vec{Q}}(t)$ is
constructed with the help of projection operator P in a
way similar to that in ref.$^{/7/}$ for two-particle case,
with $\Psi'_{\vec{Q}}(t)$ satisfying asymptotic equality (8) as the total
wave function $\Psi'_{\vec{Q}}(t)$. Eq. (9) is a relativistic generalization of the Schrödinger equation for the system with
arbitrary fixed number of particles.

Let us construct with the help of the wave function
$\Psi'_{\vec{Q}}(t)$ a density operator for a pure multiparticle state,

$$\rho_n(\vec{k}_1 \ldots \vec{k}_n, \vec{\kappa}_1 \ldots \vec{\kappa}_n | t) = \Psi'_{\underset{Q}{}}(\vec{k}_1 \ldots \vec{k}_n, t) \overset{*}{\Psi}'_{\underset{Q}{}}(\vec{\kappa}_1 \ldots \vec{\kappa}_n, t),$$

$$(10)$$

where on the RHS the wave functions are generally speaking taken at different values of energy $E \neq E'$. It is no problem to understand that the density operator introduced in such a way satisfies the equation

$$(E_k - E)\rho_n(t) - \rho_n(t)(E_\kappa - E') = V_n \rho_n(t) - \rho_n(t) \overset{\dagger}{V}_n. \qquad (11)$$

We would remark once again that the quasipotential V_n in this case is non-Hermitian and is not a sum of pair quasipotentials, in contrast with the nonrelativistic quantum mechanics.

IV. EQUATION FOR DENSITY OPERATOR
OF MIXED MULTIPARTICLE STATE

Let us for simplicity have a system consisting of one nucleon and n mesons, which arises from a two-particle collision. We will still be interested in behavior of the selected system of m mesons only (generalization for various types of particles is no problem). In this case one may introduce the density operator of a mixed state according to standard rules of quantum mechanics:

$$\rho_n^{(m)}(\vec{k}_1 \ldots \vec{k}_m; \vec{\kappa}_1 \ldots \vec{\kappa}_m | t) =$$

$$\int d\vec{k}_{m+1} \ldots d\vec{k}_n \rho_n(\vec{k}_1 \ldots \vec{k}_m \vec{k}_{m+1} \ldots \vec{k}_n; \vec{\kappa}_1 \ldots \vec{\kappa}_m \vec{k}_{m+1} \ldots \vec{k}_n | t). \quad (12)$$

If we are not interested in the total number of particles in the system then we may introduce a summed density operator

$$\rho^{(m)}(t) = \sum_{n=m}^{\infty} \frac{1}{(n-m)!} \rho_n^{(m)}(t).$$ (13)

Diagonal elements of density operator (12) are connected with inclusive cross section of two-particle collision through a simple relation. It is an attractive feature of the density operator. It is easy to show that owing to the limiting property of the wave function (8) we will have in the case of one selected particle

$$\lim_{\substack{t \to \infty \\ E'=E}} \rho^{(1)}(\vec{k}_1, \vec{k}_1 | t) = \delta(0)\frac{I}{(2\pi)^2} \frac{d\sigma}{d\vec{k}_1} ,$$ (14)

where $I = v2q_1^{\circ}2q_2^{\circ} = 4\sqrt{(q_1 q_2)^2 - \mu^2 m^2}$ and v is a relative velocity of the initial particles.

Further, it is convenient to introduce a density operator for the states where the summed momentum of nucleon and selected meson system has a value $\vec{c}$,

$$\rho_{nc}^{(m)}(\vec{k}_1 \ldots \vec{k}_m, \vec{\kappa}_1 \ldots \vec{\kappa}_m | t) =$$

$$= \int d\vec{k}_{m+1} \ldots d\vec{k}_n \delta(\vec{c} + \sum_{i=m+1}^{n} \vec{k}_i - \vec{Q})$$ (15)

$$\times \quad \rho_n(\vec{k}_1 \ldots \vec{k}_m \vec{k}_{m+1} \ldots \vec{k}_n, \vec{\kappa}_1 \ldots \vec{\kappa}_m \vec{k}_{m+1} \ldots \vec{k}_n | t).$$

It is obvious that $\rho_{nc}^{(m)}$ possesses the property,

$$\int d\vec{c} \sum_{n=m}^{\infty} \frac{1}{(n-m)!} \rho_{nc}^{(m)}(t) = \int d\vec{c} \, \rho_{c}^{(m)}(t) = \rho^{(m)}(t).$$ (16)

Let us assume that we are interested in the behavior of the first meson only; then in Eq. (11) we shall take $\vec{k}_i=\vec{\kappa}_i$ for all $i>1$, integrate out in all $\vec{k}_i (i>1)$ and sum over n, before having multiplied this equation by $1/(n-1)!$ and $\delta(\vec{c}+\sum_{i=2}^{n}\vec{k}_i-\vec{Q})$. As a result Eq. (11) turns into the equation of the following form:

$$(E_k-E)\rho_{\vec{c}}^{(1)}(t)-\rho_{\vec{c}}^{(1)}(t)(E_K-E') = \Theta(\vec{c})\rho_{\vec{c}}^{(1)}(t)-\rho_{\vec{c}}^{(1)}(t)\Theta^{+}(\vec{c}), \quad (17)$$

where $E_k=k^{o}+\sqrt{(\vec{c}-\vec{k})^2+m^2}$ and the effective quasipotential $\Theta(\vec{c})$ is determined by taking an average from the many-particle interaction quasipotential:

$$\sum_{n=1}^{\infty}(\frac{1}{n-1})! \int d\vec{k}_2 \ldots d\vec{k}_n \int d\vec{k}_1' \ldots d\vec{k}_n' \delta(\vec{c}+\sum_{i=2}^{n}\vec{k}_i-\vec{Q}) \times \quad (18)$$

$$\times \quad V_n(\vec{k}_1\ldots\vec{k}_n;\vec{k}_1'\ldots\vec{k}_n'|\vec{Q})\rho_n(\vec{k}_1^1\ldots\vec{k}_n^1;\vec{\kappa}_1\vec{k}_2\ldots\vec{k}_n|t) =$$

$$\int d\vec{k}_1'\Theta(\vec{k}_1;\vec{k}_1'|\vec{c})\rho_{\vec{c}}^{(1)}(\vec{k}_1';\vec{\kappa}_1|t).$$

Relation (17) is the required equation for the operator $\rho_{\vec{c}}^{(1)}(t)$ through which, via Eq. (16), the true density operator (13) connected with one-particle distribution through (14) is expressed.

As is seen Eq. (17) coincides in form with Eq. (5) for the density operator of the pure meson-nucleon state. However the potential $\Theta(\vec{c})$ in Eq. (17) is not equal to the meson-nucleon interaction quasipotential $V(\vec{c})$ but is screened by the presence of the remaining secondary mesons and has the meaning of a quasipotential of the selected meson interaction with the self-consistent field of all the other hadrons in the final state.

Definition of the potential $\Theta(\vec{c})$ with the help of relation (18) is certainly of a conditional character. Its calculation is very complicated problem and approximate methods cannot be avoided. However it is not difficult to establish a very important property of the quasipotential $\Theta(\vec{c})$. For this purpose we take $\vec{k}=\vec{\kappa}$ in Eq. (17) and, integrating it over $\vec{k}$ and $\vec{c}$ with the help of Eq. (14), obtain:

$$\int d\vec{c} \int d\vec{k} d\vec{k}' \, \Theta_2(\vec{k};\vec{k}'|\vec{c}) \rho_{\vec{c}}^{(1)}(\vec{k};\vec{k}|o) = -\frac{I}{(2\pi)^3}{}^{<n>}\sigma_{incl}, \qquad (19)$$

where $\Theta_2(\vec{c})$ is an anti-Hermitian part of the quasipotential $\Theta(c)$ and $<n>$ is an average multiplicity of the secondary mesons.

Thus we see that anti-Hermitian part of the screened quasipotential $\Theta(\vec{c})$ is always negative. This result can seem unexpected since all two-particle quasipotentials due to the unitarity property of the elastic scattering amplitude must have a nonnegative anti-Hermitian part that satisfies the absorption condition. However the quasipotential $\Theta(c)$ describes particle production and it is just formally connected with some fictitious two-particle process, so it is not anything surprising that this quasipotential turns out to be antiunitary.

Eq. (17) in this form coincides with the analogous equation in ordinary quantum mechanics, so as was mentioned in Section I, one may follow phenomenological construction of model quasipotential so that the solution of Eq. (17) for density operator would satisfactorily describe experimental data.

V. SOLUTION OF EQUATION FOR DENSITY OPERATOR
CALCULATION OF SINGLE-PARTICLE DISTRIBUTION

Since Eq. (17) coincides formally with Eq. (5) for
the density operator of the pure two-particle state one
may introduce a fictitious wave function $\Psi_{\vec{c}}(\vec{k};t)$ describing the system of meson and nucleon interacting with
each other through the effective antiunitary quasipotentia
$\Theta(\vec{c})$. We notice that a normalization of $\Psi_{\vec{c}}(t)$ does not
coincide with the one of the usual two-particle wave
function and must be determined by the normalization of
the whole many-particle density operator, which is
fixed by Eq. (14). So by analogy with Eq. (2) we write:

$$\rho_{\vec{c}}^{(1)}(\vec{k};\vec{\kappa}|t) = \frac{b(s)}{\sqrt{(\vec{Q}-\vec{c})^2+\mu^2}} \Psi_{\vec{c}}(k;t)\overset{*}{\Psi}_{\vec{c}}(\vec{\kappa};t), \qquad (20)$$

where we write down explicitly an invariant normalization function $b(s)$, so that the function $\Psi_{\vec{c}}(t)$ can be
considered normalized in the usual way. Later on we
shall see that a product $\Psi_{\vec{c}}\overset{*}{\Psi}_{\vec{c}}$ describes just that part
of the inclusive cross section which is determined by
the inner structure of the nucleon, and the normalization
function $b(s)$ determines the energy dependence of the
inclusive cross section which is not related with internal
structure.

Because of Eq. (17) the wave function $\Psi_{\vec{c}}(t)$ satisfies
the following quasipotential equation:

$$(E_k-E)\Psi_{\vec{c}}(t) = \Theta(\vec{c})\Psi_{\vec{c}}(t). \qquad (21)$$

To solve this equation we shall introduce for convenience
a new antiunitary quasipotential, $\tilde{\Theta}(\vec{c})$, having explicitly

separated the true potential for meson-nucleon interaction
$V(\vec{c})$ from $\Theta(\vec{c})$:

$$\tilde{\Theta}(\vec{k};\vec{k}'|\vec{c}) = \Theta(\vec{k};\vec{k}'|\vec{c}) - V(\vec{k};\vec{k}'|\vec{c}). \tag{22}$$

Now Eq. (21) has the form:

$$(E_k - V(c) - E)\Psi_{\vec{c}}(t) = \tilde{\Theta}(\vec{c})\Psi_{\vec{c}}(t). \tag{23}$$

Evidently the distortion potential $\tilde{\Theta}(\vec{c})$ is connected with
screening of the interaction between meson and nucleon
by remaining secondaries in the final state.

It needs to find the solution of the Eq. (23) within continuous energy spectrum E. For this we shall
assume that the highest discrete energy level E'_0 of
the quasipotential $\Theta(\vec{c})$ is not far from continuous
spectrum, so that $|E-E'_0| << |<\Psi_{\vec{c}}, \tilde{\Theta}(\vec{c})\Psi_{\vec{c}}>|$, where $\Psi_{\vec{c}}$ is a
wave function of the corresponding discrete state of
the quasipotential $V(\vec{c})$ with the energy E_0 (without
taking into account the screening) which satisfies an
equation,

$$(E_k - V(\vec{c}) - E_0)\psi_{\vec{c}} = 0. \tag{24}$$

It is such a situation that corresponds better to the
particle production picture which we attempt to describe
here. In the considered case the formal solution of
Eq. (23) can be written in the following way:

$$\Psi_{\vec{c}} = \chi_{\vec{c}} + \psi_{\vec{c}} + G_0\tilde{t}(\vec{c})(\chi_{\vec{c}}+\psi_{\vec{c}}), \tag{25}$$

$\tilde{t}(\vec{c})$ satisfying an equation:

$$\tilde{t} = U + U\, G_o \tilde{t}, \tag{26}$$

where

$$U = (1-VG_o)^{-1} P\tilde{\Theta} \tag{27}$$

and (I-P) is a projection operator into the described discrete state. The wave function $\chi_{\vec{c}}$ describes the system of interacting meson and nucleon and satisfies an equation:

$$(E_k - V(\vec{c}) - E)\chi_{\vec{c}} = 0. \tag{28}$$

Everywhere in the written formulae:

$$G_o = (E_k - E - ic)^{-1} \tag{29}$$

It is no problem to understand that function $\tilde{t}(\vec{k}:\vec{k}\,'|\vec{c})$ is formally the amplitude of the elastic scattering of the selected meson in the screening field of all remaining secondaries in the final state. In contrast to the real two-particle amplitude, it does not satisfy the unitarity condition because the quasi-potential $U(\vec{c})$ is antiunitary.

$$\rho_{\vec{c}}^{(1)}(t) = \frac{b(s)e^{-i(E-E')t}}{\sqrt{(\vec{Q}-\vec{c})^2+\mu^2}}\, (1+G_o\tilde{t})(\chi_{\vec{c}}+\psi_{\vec{c}})(\chi_{\vec{c}}^* + \psi_{\vec{c}}^*)(1+\tilde{t}G_o)^{++}. \tag{30}$$

This expression defines the density operator for the case when the momentum of the selected system of interacting meson and nucleon is fixed and is equal to $\vec{c}$.

In a general case, to define a density operator one needs, in accordance with formula (16), to take the superposition of all the solutions with all the possible values for $\vec{c}$:

$$\rho^{(1)}(t) = b(s)e^{-i(E-E')t}\int \frac{d\vec{c}}{\sqrt{(\vec{Q}-\vec{c})^2+\mu^2}} \times \quad (31)$$

$$\times (1+G_o\tilde{t})(\chi_{\vec{c}}+\psi_{\vec{c}})(\overset{*}{\chi}_{\vec{c}}+\overset{*}{\psi}_{\vec{c}})(1+\tilde{t}G_o)^{++} .$$

The function $\tilde{t}(\vec{k};\vec{k}'|c)$ in this expression, is known to us in the case when the screening quasipotential $\tilde{\Theta}(c)$, defined by (22) and (18) is known. We have already pointed out the phenomenological possibility of determining this potential; however it may be reconstructed with the methods used in conventional quantum mechanics when studying self-consistent field. In particular if correlations are neglected we can use a very well known Hartree-Fock method to obtaining the simplest approximation.

In concluding this Section we shall calculate an explicit expression for the single-particle distribution with the help of a limit of Eq. (14), proceeding from the obtained solution for density operator (31). In transition to the limit $t \to \infty$ in Eq. (31) and putting $\vec{k} =\vec{\kappa}$ and $E = E$; taking into account (14), we obtain:

$$\frac{d\sigma}{d\vec{k}} = \frac{(2\pi)^4}{I}b(s)\int d\vec{c}\,\delta(E_k-E)\frac{\left|\int d\vec{k}'t'(\vec{k};\vec{k}'|\vec{c})\chi_{\vec{c}}^{(o)}(\vec{k}';o) +\right.}{\sqrt{(\vec{Q}-\vec{c})^2+\mu^2}}$$

$$\left.+\int d\vec{k}'\hat{t}(\vec{k};\vec{k}|\vec{c})\psi_{\vec{c}}(\vec{k}',o)\right|^2 , \quad (32)$$

with free wave function $\chi_{\vec{c}}^{(o)}$ satisfying an equation:

$$(E_k - E)\chi_{\vec{c}}^{(o)}(k;o) = 0.$$

The first term in (32) is related to the leading effect, i.e. with the fact that in the initial state we have just a similar meson, as the selected one in the final state. This term in its structure describes an elastic scattering of the initial meson on the quasi-potential $\Theta' = V + P \tilde{\Theta}$. Therefore the amplitude $t'(\vec{c})$ satisfies the following quasipotential equation:

$$t' = \Theta' + \Theta' G_o t' \qquad (33)$$

It is worth noticing that, if from the very beginning we had considered the collision of two nucleons, then the first term would have been absent in formula (32), as the initial value for the solution of Eq. (23) had not contained the continuous spectrum.

The second term in (32) is directly connected with secondary meson production. Having analyzed the structure of this integral we come to a conclusion that the initial nucleon should be a composite system, inside which elementary mesons are distributed according to $\tilde{\psi}_{\vec{c}}(\vec{k};0)$. The amplitude $\tilde{t}(\vec{c})$ satisfies quasipotential equation (26) and describes elastic scattering on separate point-like constituents of a nucleon, taking into account the screening interaction of all the remaining constituents, because this scattering is accompanied by particle production and the amplitude $\tilde{t}(\vec{c})$ seems to be nonunitary.

Such an interpretation of the obtained results is very close in spirit to the parton idea, proposed by Feynman. It should be noted, that when there are several types of mesons then formula (32) would include a partial function of the distribution of the very type which interests us, in the final state of inclusive reaction.

VI. SCALING PROPERTIES OF THE INCLUSIVE
CROSS SECTION

Later on it will be convenient to pass on to invariant amplitudes and to wave functions in Eq. (32).
The inclusive cross section can be represented in c.m.s.
in the form:

$$2k^{\circ} \frac{d\sigma}{d\vec{k}} = \frac{(2\pi)^4 b(E^2)}{12(E-k^{\circ})} \int \frac{d\vec{c}}{2\sqrt{c^2+\mu^2}} \, \delta(E_k-E) I_{\vec{c}}(\vec{k}), \qquad (34)$$

where

$$I_{\vec{c}}(\vec{k}) = \left| \int d\vec{k}' t'^r(\vec{k};\vec{k}^1|\vec{c}) \chi_{\vec{c}}^{r(o)}(\vec{k}';o) + \right.$$

$$\left. + \int d\vec{k}' \tilde{t}^r(\vec{k};\vec{k}'|\vec{c}) \psi_{\vec{c}}^r(\vec{k};o) \right|^2 . \qquad (35)$$

Let us consider the invariant function

$$\tilde{I}_{\vec{c}}(k) = \delta(k^2-\mu^2) I_{\vec{c}}(\vec{k}). \qquad (36)$$

Let us assume that in the rest system of the selected
pair of particles, $\vec{c} = 0$; the function $I_0(\vec{k})$ is known
then in an arbitrary system of reference and one can
obtain by means of simple Lorentz transformation:

$$\tilde{I}_{\vec{c}}(k) = \tilde{I}_0(k^1) . \qquad (37)$$

The transformation connecting momenta k and k' involves
the invariant mass M of the selected meson-nucleon system.

It equals simply the mass of the discrete state described
by Eq. (24) for the second term within the modulus in
Eq. (35). In the case of the continuous spectrum:

$$M^2 = E^2 - c^2. \tag{38}$$

We shall see below that in the interesting region
of scaling behavior $k \gg \mu$, these values are of the same
order and so for simplicity we shall consider them
identical here.

As a result, it is simple to show that function $I_{\vec{c}}(\vec{k})$
has a form:

$$I_{\vec{c}}(\vec{k}) = \frac{2k^\circ M}{c^\circ} \Phi\left(\vec{k}'^2, \frac{c}{c}\circ(\vec{k}'\vec{n}) \right), \tag{39}$$

where $c^\circ = \sqrt{c^2 + M^2}$,

$$\vec{k}' = \vec{k} - (\vec{k}\vec{n})\vec{n} + \frac{M}{c}\circ(\vec{k}\vec{n})\vec{n}, \tag{40}$$

and $\vec{n}$ is a unit vector in $\vec{c}$ direction.

Substituting the expression (39) into Eq. (34) we
obtain, after replacing the integral in C;

$$2k^\circ \frac{d\sigma}{d\vec{k}} = \frac{(2\pi)^4 b(E^2) Mk}{2I} \int \frac{d\Omega_{\vec{c}}}{c-(\vec{k}\vec{n})} \times \tag{41}$$

$$\times \quad \Phi\left(\vec{k}^2 - (\vec{k}\vec{n})^2 + \frac{M^2(\vec{k}\vec{n})^2}{c^2}, \frac{M}{c}(\vec{k}\vec{n}) \right),$$

where

$$c = \sqrt{(E-k^\circ)^2 + (\vec{k}\vec{n})^2 - k^2 - m^2} + (\vec{k}\vec{n}). \tag{42}$$

In Eq. (41) we have taken into account that the initial
energy is large, $E \gg M$, and so $c \gg M$. Thus to calculate
the inclusive cross section it is necessary to integrate
over all possible directions of vector $\vec{c}$, which is the
total momentum of the selected meson-nucleon system in
the final state.

We assume that function $\Phi(\vec{k}'^2, (\vec{k}'\vec{n}))$ rapidly decreases
as $\vec{k}'^2$ is growing. Such a behavior corresponds to the
experimental observation for inclusive distributions
and apparently reflects the short range character of
nuclear forces. From the aforementioned facts it follows
that the main contribution to integral (41) is made by
the range of angles θ near θ_o where $\vec{k}'^2$ takes its
minimum. The minimum condition yields:

$$\operatorname{tg}\theta_o = \operatorname{tg}\alpha \, \cos\psi \quad , \tag{43}$$

where α is an angle of the secondary meson emission.
So we shall estimate the integral (41) removing the
integrand at the value of angle $\theta = \theta_o$

$$2k^o \frac{d\sigma}{d\vec{k}} = \frac{(2\pi)^4 b(E^2) Mx}{2E^2} \int_o^{2\pi} \frac{d\psi}{\sqrt{(2-x)^2 - x_1^2 \sin^2\psi}} \times$$

$$\Phi\left(k_1^2 \sin^2\psi + \frac{M^2 x^2}{\left(\sqrt{(2-x)^2 - x_1^2 \sin^2\psi} + \operatorname{sgn} \cos\psi \sqrt{x^2 - x_1^2 \sin^2\psi}\right)^2} \right.$$

$$\left. \frac{M \operatorname{sgn} \cos\psi \sqrt{x^2 - x_1^2 \sin^2\psi}}{\sqrt{(2-x)^2 - x_1^2 \sin^2\psi} + \operatorname{sgn} \cos\psi \sqrt{x^2 - x_1^2 \sin^2\psi}} \right) \quad , \tag{44}$$

where the Feynman variables $x = 2k/E$ and $x_\perp = 2k_\perp/E$
are introduced.

Analyzing this expression we see that the energy dependence of the inclusive cross section is factorized in the form $b(E^2)/E^2$. As was noted, such an energy dependence is not related to the inner structure of the interacting particles at high energy, but obviously it has a kinematic origin. The uncertainty of this dependence arises from the ambiguity in the normalization of the wave function of two-particle system. The remaining part of the inclusive cross section connected with the internal structure of particles depends on the transverse momentum and scaling variables $x_\perp$ and x in the region of high energies and fixed x.

In the region of the limited transverse momentum $x_\perp \ll x$, the expression (44) takes a form

$$2k^\circ \frac{d\sigma}{d\vec{k}} = \frac{(2\pi)^4 b(E^2) M |x_{\shortparallel}|}{2E^2(2-|x_{\shortparallel}|)} \times \qquad (45)$$

$$\times \int_0^{2\pi} d\psi \; \Phi\left(k_\perp^2 \sin^2\psi + \frac{M^2 x_{\shortparallel}^2}{(2-|x_{\shortparallel}|+|x_{\shortparallel}|\,\mathrm{sgn}\,\cos\psi)^2}\right.,$$

$$\left. \frac{M|x_{\shortparallel}|\,\mathrm{sgn}\,\cos\psi}{2-|x_{\shortparallel}|+|x_{\shortparallel}|\,\mathrm{sgn}\,\cos\psi}\right),$$

where $x_{\shortparallel} = 2k_{\shortparallel}/E$ and consequently depends on $k_\perp$ and scaling variable $x_{\shortparallel}$ only.

In some cases one may carry out an estimate of the integral over ψ in formula (44), removing the integrand at the point $\psi = 0$. As before this condition is dictated by requirement of the minimum of $\vec{k}'^2$. In this case the inclusive cross section takes the quite simple form

$$2k^\circ \frac{d\sigma}{d\vec{k}} = \frac{(2\pi)^5 b(E^2)Mx}{2E^2(2-x)} \quad \Phi\left(\frac{M^2 x^2}{4}, \frac{Mx}{2}\right) \qquad (46)$$

and depends on the scaling variable $x = 2k/E$ only.

It is not difficult to see that the minimum condition for $\vec{k}'^2$ in two variables θ and ψ leads to the fact that the direction of vector $\vec{c}$ coincides with the direction of $\vec{k}$. In this case it is obvious that

$$c = \sqrt{(E-k^\circ)^2 - m^2} + k \qquad (47)$$

and the invariant mass (38) for the continuous spectrum is approximately given by the expression

$$M^2 \overset{\sim}{=} \frac{2m^2}{2-x} + \frac{2\mu^2}{x}. \qquad (48)$$

Thus as we have suggested the invariant mass is a limited value and its x dependence does not violate scaling properties of the inclusive cross section.

VII. LARGE TRANSVERSE MOMENTUM REGION

In the region of large transverse momenta $k_\perp$ and limited $|k_{\shortparallel}| << k_\perp$, the formula (44) takes a form

$$2k^\circ \frac{d\sigma}{d\vec{k}} = \frac{(2\pi)^4 b(E^2)Mx_\perp}{2E^2} \int_0^{2\pi} \frac{d\psi}{\sqrt{4-4x_\perp + x_\perp^2 \cos^2\psi}} \times$$

$$\Phi\left(k_\perp^2 sn^2\psi, \frac{Mx_\perp \cos\psi}{\sqrt{4-4x_\perp + x_\perp^2 \cos^2\psi} + x_\perp \cos\psi}\right), \qquad (49)$$

and so the inclusive cross section asymptotics in $k_\perp$ is determined by the behavior of function Φ at large $k^2 \sin\psi$.

To simplify the calculations we assume that function Φ is represented in a factorized form

$$\Phi\left(\vec{k}^2, \frac{c}{c^o}(\vec{k}\vec{n})\right) = \frac{1}{2k^o} I_o(\vec{k}^2) \; f\left(\frac{c}{c^o}(\vec{k}\vec{n})\right), \qquad (50)$$

where f is an arbitrary function normalized by the condition $f(0) = 1$.

In the considered region we restrict ourselves with a calculation of the second term only inside the modulus in Eq. (35). As it was noted the first term is related to the leading particle and is reduced to elastic scattering on the given quasipotential, and its behavior at large transverse momenta is determined by corresponding properties of elastic scattering amplitude. Since $\tilde{t}^r$ is a relativistic invariant amplitude then one may consider that in the rest system of reference $c = 0$ it depends on $(\vec{k}-\vec{k}')^2$ and a total energy E. Therefore we obtain, passing on to the coordinate space,

$$I_o(\vec{k}) = (2\pi)^6 | \int d\vec{r} e^{-i\vec{k}\vec{r}} \tilde{t}^r_{E(\vec{r}|o)}(\vec{r}|o)\psi^r_o(\vec{r};o)|^2. \qquad (51)$$

Let us notice that small values of r give the main contribution to the integral in the large k limit.

For the functions $\tilde{t}^r_{E(\vec{r}|}0)$ and $\psi^r_o(\vec{r};o)$ it may be said that they should rapidly decrease with distance. Besides we have remarked that the amplitude $\tilde{t}^r_E$ describes interaction with separate constituents of the nucleon which we shall consider to be point-like. For this reason the function $\tilde{t}^r_E$ should be singular. Since in the limit of large transverse momenta interactions take place mainly at small distances, this singularity plays a

significant role in particle production. In accordance
with previous arguments, for definiteness we shall
assume that

$$\psi_o^r(\vec{r};o) = h\ e^{-2r^2},\tag{52}$$

$$\tilde{t}_E^r(\vec{r}|o) = g(E)\ \frac{e^{-\beta r^2}}{r^{2\gamma}}\ ,\quad 0<\gamma<3/2,\tag{53}$$

where h, α and β are constants and g(E) is an arbitrary
function of energy. Since amplitude $\tilde{t}_E^r$ describes elastic
hadron scattering, then on a basis of general representations for energy dependence of hadron cross sections one
may consider that up to logarithmic factors g(E) $\sim$ E^2.

   Substituting the expressions (52) and (53) into
Eq. (51) with the help of the representation (50) we
obtain:

$$2k^o\ \frac{d\sigma}{d\vec{k}} = \frac{(2\pi)^4 b(E^2)Mx_\perp}{2E^2}\ \int_o^{2\pi} \frac{d\psi}{\sqrt{k_1^2\sin^2\psi+\mu^2}\sqrt{4-4x_\perp+x_\perp^2\cos^2\psi}}\ ,\quad \times\tag{54}$$

$$\times\ f\left(\frac{Mx_\perp\cos\psi}{\sqrt{4-4x_\perp+x_\perp^2\cos^2\psi}+x_\perp\cos\psi}\right)I_o(\vec{k}_\perp^2\sin^2\psi),$$

where

$$I_o(\vec{k}^2\sin^2\psi)= \left|(2\pi)^4 hg\frac{\Gamma(3/2-\gamma)}{(\alpha+\beta)^{3/2-\gamma}}\ e^{-\frac{\vec{k}_\perp^2\sin^2\psi}{4(\alpha+\beta)}}\right.\qquad \times$$

$$\left.\times\qquad F\left(\gamma;3/2;\frac{\vec{k}_\perp^2\sin^2\psi}{4(\alpha+\beta)}\right)\right|^2.\tag{55}$$

We assume that the main contribution to the integral (54) is given by angles ψ near a fixed value $\psi_0 \neq 0$. It may be achieved let us say by means of the choice for the function f. In this case, taking the limit of large $k_\perp$ under the integral sign we obtain

$$2k^\circ \frac{d\sigma}{d\vec{k}} = \frac{(2\pi)^4 b(E^2)|g(E)|^2 M}{2E^2 (k_\perp^2)^{3-2\gamma+1/2}} \quad y(x_\perp), \qquad (56)$$

where

$$y(x_\perp) = \left| (2\pi)^4 h \, 2^{3-2\gamma} \frac{\Gamma(3/2-\gamma)\Gamma(3/2)}{\Gamma(\gamma)} \right|^2 \times \qquad (57)$$

$$\times \int \frac{d\psi(\sin^2\psi)^{-3+2\gamma+1/2}}{\sqrt{4-4x_\perp+x_\perp^2\cos^2\psi}} \, f\left(\frac{Mx_\perp\cos\psi}{\sqrt{4-4x_\perp+x_\perp^2\cos^2\psi}+x_\perp\cos\psi} \right) .$$

Here integration is carried out in the vicinity of the point $\psi_0 \neq 0$. Thus with growth of the transverse momentum the inclusive cross section decreases in the power-like form, and is dependent on scaling variable $x_\perp$. It should be noted that in this case the power form behavior is an effect caused by production at small distances which is due to the singular character of interaction between point-like constituents of the nucleon.

As $0<\gamma<3/2$, then the power index of inclusive cross section decrease is always less than 7. Elastic hadron scattering amplitudes as a rule always decrease more rapidly. Therefore an account of the first term in formula (35) would not change the asymptotic behavior of the inclusive cross section (56).

In the case when the angle ψ_0 is near zero one

cannot estimate the integral in Eq. (54) going to the
limit $k_\perp \to \infty$ under the integral sign. However it is
easy to see that in this case the inclusive cross section
dependence on $k_\perp$ will be power-like as well but with
smaller power index. Besides, it may turn out that this
index essentially depends on $x_\perp$.

VIII. CONCLUSION

In the present work in the framework of equal-time
formulation of the many-body problem in QFT a procedure
of describing inclusive reactions with the help of density
operator has been proposed. In this the density operator
is constructed of equal-time multiparticle wave functions
according to the rules of conventional quantum mechanics.
The equation satisfied with this density operator, does
not differ in its form from the corresponding equation
in quantum mechanics.

It has been shown that equation for density operator
of single-particle inclusive reactions coincides formally
with the equation for the density operator of a pure two-
particle state of meson and nucleon. However the interaction between them is distorted by the screening quasipotential, that is related to the presence of the remaining hadrons in the final state and is expressed via
two-particle and multiparticle interaction quasipotentials.
Since this potential is related to particle production,
its anti-Hermitian part seems to be negative. To calculate
the screening quasipotential one may use the same methods
as in the usual quantum mechanics when studying selfconsistent field.

The density operator constructed in this way is
connected with inclusive distribution via a simple

relation.   Therefore the solution of an equation for
the density operator allows one to express this distribution through screening quasipotential.   The result
obtained permits one to interpret the initial nucleon
as a composite system, consisting of point-like elementary
parts, distributed according to a certain law, and the
production of secondary meson as scattering separate
point-like constituents accompanied by knocking them out.

    In Sections VI and VII it has been shown that some
model assumptions on screening quasipotential allows one
to reveal a scaling behavior of the inclusive cross
section and power form decrease with transverse momentum
growth.

    The authors express their deep gratitude to M. A.
Mestvirishvili, M. V. Saveliev, N. E. Tyurin and
O. A. Khrustalev for fruitful discussions.

REFERENCES

1. A. A. Logunov, A. N. Tavkhelidze, Nuovo Cim., $\underline{29}$, 380, 1963.

2. A. A. Logunov, V. I. Savrin, N. E. Tyurin, O. A. Khrustalev, TMF, $\underline{6}$, 157, 1971.

3. V. R. Garsevanishvili, V. A. Matveev, L. A. Slepchenko, A. N. Tavkhelidze. Phys. Lett., $\underline{29B}$, 191. 1969.

4. V. A. Matveev, R. M. Muradian, A. N. Tavkhelidze. Preprint JINR P2-3900, Dubna, 1968; A. N. Kvinikhidze, D. Ts. Stoyanov. TMF, $\underline{3}$, 332, 1970; V. M. Vinogradov. TMF, $\underline{8}$, 343, 1971.

5. A. A. Arkhipov, V. I. Savrin. TMF, $\underline{16}$, 328, 1973; TMF, $\underline{19}$, 320, 1974.

6. K. J. Biebl, J. Wolf. Nucl. Phys., $\underline{B44}$, 301, 1972; J. C. Botke, D. J. Scalapino, R. L. Sugar. Phys. Rev., $\underline{D9}$, 813, 1974.

7. V. I. Savrin, N. E. Tyurin. Preprint IHEP STF 73-31, Serpukhov, 1973.

MY PERSPECTIVES ON PARTICLE PHYSICS:

SUMMARY OF ORBIS SCIENTIAE, 1976

Benjamin W. Lee

Fermi National Accelerator Laboratory

Batavia, Illinois 60510

TABLE OF CONTENTS

I.   INTRODUCTION - EXCUSES AND APOLOGIES

When my old teacher and friend, Syd Meshkov, called me some weeks ago to ask if I would summarize the conference, I happened to be rereading a letter Vicki Weisskopf received from his teacher at the time of the discovery of parity violation -- another era of excitement and rapid progress in particle physics. It was fascinating reading, because the writer was musing on many of the problems that we are trying to unravel and understand even today. But I am not here to elaborate on Pauli's insight; what inspired me to accept this challenging task was the last sentence in this historical document,[1] which was

"Viele Fragen, keine Antworten."

This is, in the summer talk I don't have to answer any profound questions, but I need only summarize questions raised in the conference! This is what gave me the courage to accept the challenge.

It would be dishonest if one were to attempt to summarize talks and discussions one did not understand. I know my shortcomings, and will have to pass unmentioned some topics. I apologize to those concerned for this unavoidable slight.

Lastly, I must quote Dick Feynman[2] verbatim: "As is conventional in summary talks, I'm not going to summarize what happened during today, the last day of the conference. The excuse usually is, of course, that you've just heard it. But I have just discovered the real reason, for the summarizer didn't attend today's sessions, because he [was] busy preparing his summary."

II.   ULTIMATE THEORY - IMPOSSIBLE DREAM?

It seems to me that an ultimate theory should

encompass all known interactions, gravitational as well
as strong, weak and electromagnetic interactions. I
do not know much about supergauge theories developed and
explained here by Peter Freund, nor a different formulation by Arnowitt, Nath and Zumino,[3] but I consider
them as an important first step toward such ambitious
undertakings. More immediately, scalar mesons some
people seem to want may find a natural setting in such
a theory.

An ultimate theory would have only one dimensional
constant, the Planck mass, and perhaps a cosmological
constant. All other physical quantities should be
computable. It must explain all natural phenomena --
from galaxies to quarks and beyond.

A revision of quantum mechanics may also be
necessary. As my mentor Chen Ning Yang[4] often emphasized
to me, an essential difference between classical and
quantum mechanics is that the underlying algebras of the
two mechanics are real and complex, respectively. The
view that one should be able to choose the phase of a
complex wave function independently at every space-time
point then naturally leads us to the concept of gauge
fields -- a single overwhelming theme of this conference.

Are there other algebras that could be, or should
be brought into physics?[4] Are there other symmetries
(other than those associated with complex algebra:
time reversal/charge conjugation and U(1) gauge invariance) that may be due to more complex underlying
algebraic structures?[4] Feza Gürsey, I believe, has
made an important observation in this connection, by
finding a natural setting for the perfect color SU(3)
symmetry in the octonion quantum theory. We will discuss this matter later.

III.   PATI-SALAM THEORY - A VIABLE ALTERNATIVE

Pati made an impassioned case for a unified theory
of leptons and integrally charged Han-Nambu quarks.
The Pati-Salam model is a viable alternative to the
more popularly held quantum chromodynamics. It is
consistent with all known experimental facts.

If it turns out theoretically that color cannot be
confined, and if fractionally charged quarks are not to
be found experimentally, then the Pati-Salam model will
be in a much stronger position as the unified gauge
theory of strong, weak and electromagnetic interactions.
If the lifetimes of quarks are of order $10^{-11} \sim 10^{-12}$
sec., as Pati estimates, emulsion exposures may prove
to be the most rewarding hunting grounds for live quarks.

Frankly, I like quantum chromodynamics (QCD hereafter) better, because it seems to me richer in mathematical beauty and complexity, and more intricate in its
physical implications. Here, though, I must admit that
I am guilty of scientific prejudice; for most of us,
it's a matter of taste, and fashion. The following discussions are mostly in the framework of QCD.

IV.   QUANTUM CHROMODYNAMICS - CENTRAL DOGMA

Murray Gell-Mann presented a beautiful introduction
to this subject and I need only list a few important
features (central dogma) of the theory:

(1). Strong, weak and electromagnetic interactions
are described by a non-Abelian gauge theory based on a
simple group G. Therefore, there is only one gauge
coupling constant. The super group G decomposes into*

_____

*The super symmetry, or supersymmetry G hereafter has
 nothing to do with the supergauge theories discussed
 in the last section.

$$G \supset SU_c(3) \otimes G_{flavor}$$

$$\supset SU_c(3) \otimes U_W(2)$$

where $SU_c(3)$ is the exact $SU(3)$ color symmetry associated with strong interactions, which can be described
by an asymptotically free gauge theory (in the narrow
sense, this is QCD), and $U_W(2)$ is the spontaneously
broken symmetry of weak (and electromagnetic) interactions.

    (2). Main ingredients of the theory are:
        Gauge bosons:
            a. confined color gluons.
            b. weak bosons and photon; other massive
               bosons which carry flavor.
            c. diquarks/leptoquarks which carry both
               color and flavor.
        Fermions:
            Leptons, and confined quarks. They appear
            in common multiplets of the super group G.
        In addition there may be fundamental scalars
        responsible for spontaneous breakdown of G
        down to $SU_c(3) \otimes U_W(2)$ and/or of $U_W(2)$.

In this general framework there are a number of
questions that have been discussed in this conference.
I will again list them, with the names of speakers who
addressed each question:

    (1). What is the supergroup G? (Gürsey)
    (2). How is color confined? (Appelquist, Cornwall,
        Wilson)
    (3). Leptonic interactions of hadrons. (De Rujula,
        Politzer)
    (4). Hadronic weak interactions. (Minkowski, Zee)
    (5). Phenomenological description of hadrons as

confined quarks and gluons in a "bag" (Johnson).   I
must stress here that the MIT bag model was first con-
ceived as a fundamental theory of hadrons.   I am taking
the liberty of interpreting it as a reasonable phenomenological description of confined quarks and gluons
in an asymptotically free QCD.

(6). Higgs meson phenomenology.

(7). Magnetic monopole (Dirac, Goldhaber, Hagen).
Dr. Hagen clarified for us the difficulties in the
operator formulation of magnetic monopole theory.

(8). "Stability of the Proton" (Gell-Mann).

(9). The origin of quark masses, Cabibbo angle and
CP violation (Glashow, Zee).

(10). Solitons (Neveu).   The relevance of quantum
solitons in QCD is not enirely clear to me.   The theory
of solitons is an interesting subject on its own right;
important progress is being made here, and our general
understanding of quantum field theory is deepened thereby.

Before summarizing my understanding of some of
these topics, let us list the experimental discoveries
that have been made during the past year or so at a
breathtaking rate.

V.   EXPERIMENTAL DEVELOPMENTS - FACTS?

(1). Discoveries of J, $\psi(3.7)$, and the physics of
the psion family, including possibly five resonances in
the range 4.0 - 4.5 GeV; μe events as possible signal of
heavy lepton production (Chen, Oberlach, Tanenbaum;
Gilman).

(2). Structure of the neutral current (Sciulli;
Minkowski).

(3). Dimuon effects in ν- and $\bar{\nu}$-induced reactions

(Mann; De Rujula).

   (4). Production of large $p_{\perp}$ leptons,[5] and $\ell/\pi \sim 10^{-4}$.

   (5).. Apparent $\Delta Q = -\Delta S$ neutrino interaction at BNL.[6]

   (6). $Ke^{+}\mu^{-}$ events in ν-induced reactions in Gargamelle and at Fermilab.[7]

   (7). High y and low x anomaly in $\bar{\nu}$-(and possibly ν-) induced reactions (Mann; De Rujula).

   (8). Scaling breakdown in μp scattering.[8]

   (9). ISR production of Kp, πp resonances (Sassoms).

   (10). Magnetic monopole (Price, Ross) - my impression on this is that there is a candidate, but the evidence is far from conclusive.

   Let me summarize items (1), (2), (3) and (7) above that have been discussed at this conference.

               VI.   PSION PHYSICS - PUZZLES

   We have heard excellent reviews from three ex-perimentalists and a theorist. It is silly for me to try to summarize the vast amount of data and possible inferences. I will instead list my own inferences and puzzles.

   (1). The spectrum of the psion family is un-mistakably that of a $Q\bar{Q}$ system, where Q is a spin $\frac{1}{2}$ fermion.

   (2). The electric charge of Q is not known: the VMD phenomenology based on SU(4) is ambiguous, because we do not know to what quantity $[1/\gamma_V$, or $(M_V)^N/\gamma_V$, where $1/\gamma_V$ is the photon-vector meson coupling and N is some integer, say] SU(4) should be applied; in-ferences based on the saturation of a Weinberg sum rule ought to take into account not just J(3.1), but ψ(3.7) and resonances in the range 4.0 - 4.5 GeV.

(3). Radiative decay widths of the P states, and of
ψ' into the P states are not in accord with charmonium
estimates. It may indicate that the charmonium wave
functions used are not good enough, and/or the charge
of Q is -1/3, rather than 2/3, in which case theoretical
estimates of E1 moments should be reduced by a factor
of 4. In any case, if the psion family can be described
in a nonrelativistic approximation, the dipole sum rule
ought to hold:

$$\sum_f \left| <i|\vec{x}|f> \right|^2 (E_f - E_i) = \frac{2}{m_Q} \quad .$$

This should be used extensively as a diagnostic tool to
determine what the charge of Q is, and to see whether
there should be large E1 transition rates from the 4.0
- 4.5 GeV region and continuium states to the P states.

(4). 20 - 25% of decay modes of ψ' are unaccounted
for. There are two possibilities again: some or all
branching ratios are underestimated, and/or there are
undiscovered decay modes, such as $\psi' \rightarrow X'(2^1S_0) + \gamma$.
In the latter case, which is an M1 transition, Q may
have a large anomalous magnetic moment.

(5). Charm searches at SPEAR and DORIS have so far
yielded negative results. Why? Let us speculate.

(5.a). Psions are bound $(b\bar{b})$ states where $Q_b = -1/3$.
The "bottom" quark decays nonleptonically by the scheme

$$b \rightarrow u\bar{u}d \quad ,$$

and there will be no "strangeness-signal" in the final
states of the $(b\bar{q})$, $(\bar{b}q)$ meson decays, where q is a
light quark, u or d.

(5.b). The usual charm scheme is right as to
charmed meson production in e^+e^- collisions, for example:

$$e^+e^- \rightarrow \gamma \rightarrow D^+D^- \quad ,$$

but D mesons decay predominantly into a heavy lepton[9]
($\pi\mu$ puzzle all over again!):

$$D^+ \rightarrow U^+ + \nu_U$$

$$\hookrightarrow \mu^+ \nu_\mu \bar{\nu}_U$$

$$\hookrightarrow e^+ \nu_e \bar{\nu}_U$$

$$\hookrightarrow \bar{\nu}_U + \text{mostly nonstrange hadrons.}$$

This hypothesis is viable, I believe,[10] if the D meson
decay constant f_D is much larger[11] than $f_\pi \approx f$
($f_D \approx 10f_\pi$, say) and $M(U) \approx 1.8$ GeV.

(5.c). The charm quark decays mostly into a u-quark
with emission of a gluon.[12] The D^+, D^o mesons decay
mostly into nonstrange hadrons.

(5.d). There is nothing wrong with the orthodox
(i.e., conventional[13]) charm phenomenology. SPEAR and
DORIS have been very, very unlucky.

Lest any of you misunderstand, let me say that I
am not advocating that the psions are bound states of
the bottom quark and its anti-quark. Rather, I am
provoking you to come up with a plausible indication
that the charge of Q is indeed 2/3.

VII. NEUTRAL CURRENT - A MILD SURPRISE
Frank Sciulli presented an analysis of the spin

structure of the weak neutral current based on the
Caltech-Fermilab experiment. His group used the Fermilab dichromatic beam, and consequently had some control
over incident neutrino energy.

Instead of reviewing the analysis, let me summarize
their result in the framework of the minimal (Weinberg-
Salam) theory of weak interactions. We can parametrize
the neutral current interactions of hadrons in an
effective Lagrangian:

$$\frac{G_F}{\sqrt{2}} \; x \; \bar{\nu}\gamma_\mu(1-\gamma_5)\nu[V_3^\mu - A_3^\mu - 2\sin\Theta_W j_{e.m.}^\mu] \; .$$

What his group found is that

$$x \approx 1 \; ,$$

$$\sin^2\Theta_W \approx 0.3 \sim 0.4 \; .$$

I would like to call your attention to the parameter x. It is a measure of the relative strength of
neutral current to charged ones; it takes the value

$$x = \frac{m_W^2}{m_Z^2 \cos^2\Theta_W} = 1$$

only in the most naive version of the theory in which
the spontaneous symmetry breakdown occurs in the doublet
Higgs system. That this prediction is borne out is a
mild surprise to me, because I expected the pattern of
spontaneously broken symmetry to be more complicated,
but perhaps not to Steve Weinberg and Abdus Salam and
to most of you. There is a lesson for me somewhere here;
I know not what, though.

VIII.   DIMUON EVENTS AND HIGH y, LOW x ANOMALY
        "Peripheral Collisions Just Study
        Fluffs. Lepton Scattering Gets at
        The Guts." -usually attributed to
        R.R. Wilson

Al Mann reported on two results of the Harvard-
Pennsylvania-Wisconsin-Fermilab neutrino experiments.

(1). Dimuon Events. Through extensive discussions
with members of the collaboration at Fermilab, I am
convinced that dimuon events are genuine, in the sense
that second muons in most of these events are not from
π/K decays, and they are not due to accidental coinci-
dences. Further, the Pais-Treiman test shows that most
of them do not come from the production and decay of a
neutral heavy lepton, $L^{\circ} \rightarrow \mu^{-}\mu^{+}\nu$.

Basic characteristics of these events are: for in-
cident ν, μ^{-} is almost always fast, μ^{+} slow; mutatis
mutandis for incident $\bar{\nu}$;

$$\frac{\sigma(\nu \rightarrow \mu^{-}\mu^{+})}{\sigma(\nu \rightarrow \mu^{-})} \approx 10^{-2}, \quad \frac{\sigma(\bar{\nu} \rightarrow \mu^{+}\mu^{-})}{\sigma(\nu \rightarrow \mu^{-}\mu^{+})} \approx 0.8 \pm 0.6$$

and

$$\frac{\sigma(\nu \rightarrow \mu^{-}\mu^{-})}{\sigma(\nu \rightarrow \mu^{-}\mu^{+})} \approx 10^{-1}$$

(2). High y and low x anomaly: Al Mann elaborated
on the high y anomaly in the deep inelastic $\bar{\nu}$ scattering
data that they have accumulated for some time; the
phenomenon is most pronounced for $E_{\bar{\nu}} > 70$ GeV, although
the trend is already apparent in 30 GeV $< E_{\bar{\nu}} < 70$ GeV.
I will present below schematic diagrams (a theorist's
conception) of their $d^{2}\sigma/dxdy$ for E_{ν} and $E_{\bar{\nu}} > 70$ GeV:

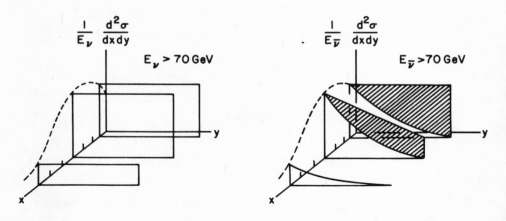

In the antineutrino case, the crosshatched portions are
excesses over the naive parton model expectation. The
excesses are concentrated in small x and large y. On
the other hand, the neutrino data seem, <u>more or less</u> to
agree with the parton model expectation. A more dramatic
demonstration of the anomaly is to plot the average value
of y, $\langle y \rangle_{\bar{\nu}}$, against $E_{\bar{\nu}}$:

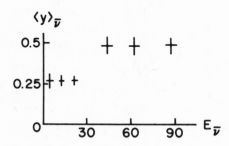

Again, the plot is my impression of the data Mann

presented; for the real data, please refer to him. The
naive (valence) parton model expectation is $1/4$, which
agrees with the low energy data. The effective threshold
for the anomaly appears to be somewhere between 30 and
50 GeV.

I shall postpone possible theoretical interpreta-
tions of these phenomena until we have examined various
models for flavors.

IX.   MODELS OF WEAK INTERACTIONS - HOW MANY FLAVORS?

With the existence of neutral currents now generally accepted, the most popular gauge theory of weak and
electromagnetic interactions is that based on the group
$SU(2) \otimes U(1)$. There are two models for fermions within
this framework that we heard about at this conference.

The minimal scheme is the original proposal of
Weinberg and Salam. In this scheme we have four doublets
of lefthanded fermions:

$$\begin{pmatrix} u \\ d_c \end{pmatrix}_L \quad \begin{pmatrix} c \\ s_c \end{pmatrix} \quad \begin{pmatrix} \nu_e \\ e \end{pmatrix} \quad \begin{pmatrix} \nu_\mu \\ \mu \end{pmatrix}$$

Another scheme which has become popular recently
is the six-quark, vectorlike model, in which we have,
for quarks:

$$\begin{pmatrix} u \\ d_c \end{pmatrix}_L, \quad \begin{pmatrix} c \\ s_c \end{pmatrix}_L, \quad \begin{pmatrix} t \\ b \end{pmatrix}_L : \quad \begin{pmatrix} u \\ b \end{pmatrix}_R, \quad \begin{pmatrix} c \\ s \end{pmatrix}_R, \quad \begin{pmatrix} t \\ d \end{pmatrix}_R$$

and for leptons,

$$\begin{pmatrix} \nu \\ e \end{pmatrix}_L, \quad \begin{pmatrix} \nu^{\prime} \\ \mu \end{pmatrix}_L, \quad \begin{pmatrix} E^{\circ} \\ U^{-} \end{pmatrix}_L \quad : \quad \begin{pmatrix} E^{\circ} \\ e \end{pmatrix}_R, \quad \begin{pmatrix} M \\ \mu \end{pmatrix}_R, \quad \begin{pmatrix} \nu_U \\ U^{-} \end{pmatrix}_R$$

where $M = \bar{M}$, i.e., M is a massive Majorana neutral. The phenomenology of leptons in this scheme was discussed by Peter Minkowski. I might add a comment here: someday we may want to look for neutrino-less double μ decays of charmed particles, such as $D^{+} \rightarrow K^{-} + \mu^{+} + \mu^{+}$, if this model hasn't been demolished by then.

The vectorlike model above is unique, to within small admixing, if we insist on the smallest number of fermions, and on the quark-lepton correspondence, and if we heed the phenomenological requirements that both $\Delta S = 0$ and $\Delta S = 1$ β decays are V-A, and the relative phases of the $\Delta I = \frac{1}{2}$ and $\Delta I = 3/2$ amplitudes in $K \rightarrow 2\pi$ and $K \rightarrow 3\pi$ agree with experiment.[14]

There are several motivations for the vectorlike theory; I have discussed them at the SLAC Conference at some length, but let me briefly repeat them here.

(1). Aesthetics: A vectorlike theory is asymptotically symmetric with respect to L↔R, asymptotically in the sense of neglecting fermion masses. Consequently, such a theory is automatically anomaly-free. Further, such a theory affords the possibility of interrelating or "deriving" CP violation, the Cabibbo angle and fermion masses in a natural way. Shelly Glashow gave an example of the first two (i.e., the Cabibbo angle and CP violation); Tony Zee the last two. (As Glashow mentioned, it is also possible, as shown by Pakvasa and Sugawara,[15] and more recently by Maiani,[16] to introduce CP violation "naturally" in a six-quark model which makes use only of V-A currents).

(2). Octet enhancement, some proponents argue, comes out more naturally in the six-quark model. Further, the "embarassment" of what I shall call the "P_6 problem", namely that $K° \to \pi\pi$ is forbidden in the Su(3) limit in conventional models (and in the minimal model) does not arise in this vectorlike model. I shall come back to these points presently.

It is worth emphasizing that in any vectorlike model, the hadronic, electronic and muonic neutral currents are always vector (therefore there are no parity violating effects in atomic physics). On the face of it, this seems to contradict Sciulli's report. However, proponents of the vectorlike theory in the audience, take heart! His results are only two standard deviations away from a pure vector theory.

X. $\Delta I = \frac{1}{2}$ RULE AND ALL THAT

As Cabibbo and Gell-Mann pointed out many years ago, the parity-violating octet part of the hadronic weak inter-actions in V-A theory transforms like λ_6, and consequently, the decay $K° \to \pi\pi$ is forbidden in the SU(3) limit. This is the P_6 problem. I personally do not consider it a problem. In all dynamical models I have examined, the suppression factor due to this is typically $(m_K^2 - m_\pi^2)/m_K^2$; it is zero in the SU(3) limit, but in reality it is about 1. If the strangeness-changing nonleptonic weak inter-actions arise from the product of a V-A and a V+A current, the parity violating octet part transforms like λ_7, and it is true that there is no "no-go" theorem even in the symmetry limit to worry about.

There are several explanations which have been ad-vanced recently for the observed, approximate $\Delta I = \frac{1}{2}$ rule, and more generally the octet enhancement rule. Let me review them.

(1). In the minimal theory, it was observed that the short distance behavior of the product of two V-A currents

is more singular for an octet part[17] than for the 27 part
if the effects of strong interactions are described by
QCD. The weak transition amplitude is described by

$$\int [d^4x]_E \ \frac{e^{-m_W r}}{r^2} \ <f|T(j_\mu(it,\vec{x})j^{\dagger\mu}(0))|i> \ , r^2 = x^2 + t^2 . \quad (1)$$

What matters therefore is the size of the matrix element
at short distances of order $1/m_W$. Using the Wilson
operator product expansion, one finds

$$T(j_\mu(x)j^{\dagger\mu}(0)) \approx |\log \mu r|^\alpha 0 + |\log \mu r|^{\alpha'} 0'$$

$$\text{for} \quad \mu r << 1 \quad\quad\quad\quad (2)$$

where the operator 0 is an octet and $0'$ is a mixture of
8 and 27; μ is a scale parameter. The exponents α and
α' are computable in QCD: with four flavors and three
colors $\alpha = 0.48$, $\alpha' = -.24$. Murray Gell-Mann and the
Caltech group consider this short distance enhancement
negligible; Mary K. Gaillard and I stated[17] that the
short-distance enhancement was not big enough. We felt
that there had to be additional enhancement of matrix
elements of 0 and suppression of matrix elements of $0'$.
I can now make this statement a little more quantitative:
for finite r we can estimate the size of the matrix element
$<f|T(j_\mu(x)j^{\dagger\mu}(0))|i>$ by inserting a complete set of intermediate states between the two currents. For r of order
$(1 \text{ GeV})^{-1}$, only a well-defined set of states makes important contributions, and these contributions are mostly octet. As r decreases, the matrix element of the lefthand side of (2) must match that of the right-hand side.
Thus the following picture emerges:

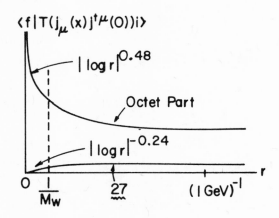

Even though my collaborators and I have not completed the detailed computation, I am not convinced yet that this picture, the cooperative enhancement of the octet channel both at long- and short-distances, cannot explain the observed validity of the $\Delta I = \frac{1}{2}$ rule.

(2). Tony Zee reported on the work of the Princeton group on this issue. The vectorlike theory contains the piece that looks like, in the local limit,

$$\sim \frac{G_F}{\sqrt{2}} \sin \Theta_c : \bar{s}_R \gamma_\mu c_R \bar{c}_L \gamma^\mu d_L : \ . \tag{3}$$

This is the product of a V-A and a V+A current; it transforms like λ_6 for p.c., and λ_7 for p.v. part. In QCD, the above interaction is multiplied (enhanced) by a factor $(\log \frac{M_W}{\mu})^\alpha$. But here α is bigger than in the minimal theory: as a rule α is bigger for (V+A)×(V-A) than for (V-A)×(V-A); the more flavors, the bigger α (up to a point). On the other hand, they argue, and I agree, that the operator in (3) will not have matrix elements of

appreciable size. It is, however, just a feeling, in-
tuition you might say, based on a naive quark model (i.e.,
there are not many c-quarks in a hadron.) Note, however,
that the product of the two V-A currents discussed in the
last section does get larger enhancement from a larger
α in a vectorlike theory.

     (3). Peter Minkowski presented a new idea on the
subject based on his work in collaboration with Harold
Fritzsch. The idea is a very attractive one. It is best
expressed by a Feynman diagram:

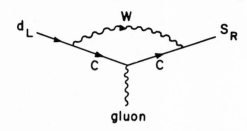

that is, the weak interaction takes place effectively
through s → d + gluon:

$$\sim \frac{G_F}{\sqrt{2}} \sin \Theta_c m_c \log\left(\frac{m_W}{m_c}\right) \bar{s} \ \sigma_{\mu\nu} \lambda_c (1-\gamma_5) d \cdot G^{\mu\nu}_{\sim} + \text{h.c. ;}$$

(4)

it has the same SU(3) transformation properties as (3).
As far as I can tell, the matrix element $<f|\bar{s}_R \sigma_{\mu\nu}\lambda_c d_L \cdot G^{\mu\nu}_{\sim} \times |i>$ will have an appreciable size. While this model is
much too new for me to give it a clean bill of health, I
do not see anything wrong so far. I do not think that this
model is grossly in conflict with the small $K_L K_S$ mass
difference.

     We have discussed two models whose structure of
effective nonleptonic weak interactions is of the form
$S_6 + P_7$. As emphasized both by Minkowski and Zee, radia-
tive decays of hyperons $(\Xi^- \to \Sigma^-\gamma, \Xi^\circ \to \Lambda\gamma, \Sigma^+ \to p\gamma)$ will prove

to be testing grounds for P_6 vs. P_7. Theoretical expectations[18] on asymmetry parameters in these decays based on $S_6 + P_6$ have not so far been borne out by experiments of limited statistics.

XI.   DIMUON EVENTS -SIGN OF CHARM?

As we have seen, the theory of nonleptonic weak decays is in a state of flux. Even in the minimal theory it is possible that the initial estimates[13] of nonleptonic branching ratios of charmed particles are in error, and I am now persuaded[19,20] that the inclusive muon branching ratio of a generic charmed particle could be as much as 20%.

The dimuon events can be explained, then, in the framework of the minimal theory.[21] They can also be explained in the vectorlike model discussed above, with somewhat different conclusions, but let me just concentrate on the former. (For the latter, please recall Alvaro DeRujula's discussion).

In the parton model, in addition to the valence quarks, there are strange quarks and their antiquarks in the sea, their contributions to the F_2 function being, say, about 5%. There are three processes that can produce charmed quarks in the final states:

1. $\nu + d \rightarrow \mu^- + c$ (Cabibbo-suppressed, but off valence and sea quarks),

2. $\nu + s \rightarrow \mu^- + c$ (Cabibbo-favored, but off sea quarks),

3. $\bar{\nu} + \bar{s} \rightarrow \mu^+ + \bar{c}$ (Cabibbo-favored, but off sea quarks).

The processes (1) and (2), summed together, would contri-

bute about 10% to the total ν cross-section well above
threshold. The process (3) would also contribute 15% to
the total $\bar{\nu}$ cross-section: recall that $\sigma_{\bar{\nu}}/\sigma_{\nu}$ for non-
charm production reactions is about 1/3.

The x distribution of the charm producing ν-reactions
will have two components:[22] one concentrated about x = 0,
representing charm production off sea quarks; the other
having a normal x distribution, representing charm production off valence quarks. In the $\bar{\nu}$ case, there is only
the small-x component. The y distribution is flat in
either case, except near y=0, where threshold effects may
manifest themselves. With the generic branching ratio
into μ channels of order 10%, one estimates

$$\frac{\sigma(\nu \to \mu^- \mu^+)}{\sigma(\nu \to \mu^-)} \approx \frac{\sigma(\bar{\nu} \to \mu^+ \mu^-)}{\sigma(\bar{\nu} \to \mu^+)} \approx 1 \sim 2\% \; .$$

The dimuon events of the same sign are problematical,
but not too serious. Charmed pair production of about 1%
of the total cross-section both in ν and $\bar{\nu}$ induced reactions would explain them comfortably. As a comparison,
I recall for you that strange pair production in the $\bar{\nu}$
interactions might be as high as 20%.[23] In the vectorlike
model, there is the additional possibility of $D^\circ \bar{D}^\circ$ mixing,
producing dimuon events of the same sign.

XII.   NEW PARTICLE PRODUCTION AND SLOW SCALING

For the following discussion, I shall assume Mann's
data on high y, low x anomaly are quantitatively correct.
My trusted colleague, Robert Shrock, and I have pondered
over the data for the last month or two, and we have ten-
tatively concluded that they cannot be understood by charm
production and logarithmic scaling breaking that QCD

predicts, alone; in particular, the jump in $\langle y \rangle_{\bar{\nu}}$ they observe is about twice (or even thrice, according to the experimenters) bigger than we can account for.

Alvaro DeRujula reported on, I believe, Michael Barnett's work on the interpretation of this anomaly in the framework of the vectorlike theory and slow scaling; to the latter David Politzer addressed himself very eloquently. But before embarking on that, let me take the 6 quark vectorlike model, and derive the most naive expectations for new particle production in ν- and $\bar{\nu}$-induced reactions, in order to illustrate the necessity of slow approach to scaling.

In the vectorlike model, there are two new elementary processes for new particle production which are Cabibbo-favored and which take place off valence quarks:

$$\nu + d_R \rightarrow \mu^- + t_R ,$$

$$\bar{\nu} + u_R \rightarrow \mu^+ + b_R .$$

The other processes are either Cabibbo suppressed or off sea quarks, and I shall ignore them. To simplify our argument, we shall assume $m(t) > m(b)$, so that the effective threshold for the former has not been reached. The second process, which takes place only in the $\bar{\nu}$ case, will have a flat y distribution. We then expect $\langle y \rangle_{\bar{\nu}}$ to jump from 1/4 to 7/16 at the effective threshold. This is an excellent agreement with observation.

However, there are two problems in this naive approach. One is that the x distribution of the new particle production in $\bar{\nu}$-induced reactions is expected to be normal; we have been told that it peaks near x=0. The second, and equally serious problem is that $\sigma_{\bar{\nu}}/\sigma_{\nu}$, which is 1/3 at low

energies, must jump to 4/3 above the effective threshold
if b alone is excited, and to 1 if both t and b are excited; at the currently available highest energy, this
ratio is at most 1/2.

Politzer described his work with Howard Georgi.
First he described what he means by the mass of a confined
(and therefore asymptotically inaccessible) quark.
Second, he discussed the operator product expansion which
takes into account the finite quark masses to all orders,
but the gluon coupling constant g to lowest order. This
study culminates in a new scaling law:

$$\nu W_2(\nu, q^2) = F_2(\xi)$$

where the variable ξ is a rather complicated function of
ν, q^2 and the initial and final quark masses m_I and m_F.
I shall not write it down here; suffice it to say that in
the case $m_I = m_F = 0$, it reduces to the usual scaling
variable x. What is the meaning of ξ? Feza Gürsey tells
me that he and Orfanides noticed some years ago that the
Clebsch-Gordan coefficients of the Poincaré group become
functions of the scaling variable ξ alone in the limit
$q^2 \rightarrow -\infty$.

For our purpose, the most relevent case is $m_I \tilde{~} 0$ and
m_F large (in the scale of hadron physics: 1 GeV, say),
which is the case considered by Barnett. In this case

$$\xi \approx x + \frac{m_F^2}{2mE_\nu y} ,$$

which can be readily understood in the parton model. Note
that x and y are restricted by the condition $W^2 = 2mEy(1-x) >$
W^2 threshold.

According to Barnett's analysis, which I saw for the

first time here, the x distribution of new particle production is squashed toward x = 0, because x $\lesssim$ ξ; the
approach to the asymptotic value of $<y>_{\bar{\nu}}$ as a function of
$E_{\bar{\nu}}$ is not very slow; the approach to the asymptotic value
of $\sigma_{\bar{\nu}}/\sigma_{\nu}$ is very slow (I must confess that I do not understand this difference intuitively at the moment; it must
be that it requires a numerical computation to reproduce
it).

So much for phenomenology. Let me now turn to more
abstract aspects of theoretical physics.

XIII. SUPER SYMMETRY - EXCEPTIONAL GROUPS AND OCTONIONS

The octonion algebra is a noncommutative, nonassocia-
tive division algebra (i.e. ax + b = 0 implies a unique
x), with seven imaginary units e_i^2 = -1, i = 1 to 7.
Suffice it to say that the e's are closed under multiplication. The automorphism group of the octonion algebra
is G_2, the first exceptional group, just as it is SU(2)
for the quarternion algebra.

Feza Gürsey's general philosophy on octonion quantum
theory is eloquently described in his Baltimore lectures.[24]
Octonion quantum theory is not a closed subject - it still
needs to be developed. The novel feature in Gürsey's
philosophy is that say, e_7 has to be used to describe
temporal development of a quantum mechanical system, i.e.,
$e^{-iEt} \rightarrow e^{-e_7 Et}$, and the automorphism group relevant to the
internal symmetry of fundamental fields is a subgroup of
G_2 which leaves $(1,e_7)$ invariant -- $SU_c(3)$! Only the color
singlet operators correspond to observables. Colored
quarks, for example, lie in a fictitious octonionic Hilbert
space; the physical Hilbert space, which is a subspace,
consists of color-neutral states. The crucial question,
to me, is then whether the physical Hilbert space is a

subspace of the bigger fictitious Hilbert space which is
automatically invariant under the action of the S-matrix
by the very nature of as-yet-to-be-formulated octonion
quantum field theories, or this must be arranged by a
special dynamical mechanism.

In any case, let our fancy take a flight! It turns
out that 3×3 hermitian matrices with octonionic entries
(more technically, octonions over real, complex quartenion
and octonion algebras, i.e., Rozenfeld algebras) are power
associative (i.e., $A^2A = AA^2 = A^3$, etc.), and satisfy the
properties of observable density matrices. The automorphism groups of these matrices comprise the rest of
exceptional simple Lie groups, F_4, E_6, E_7 and E_8. Gürsey
proposed to consider these groups as candidates for the
super unified group G.

Much work has been done along these lines by Gürsey
and collaborators, and by Gell-Mann. A remarkable feature
of all these groups is that they contain a maximal subgroup
$SU_c(3) \otimes G_{flavor}$: for E_6, E_7 and E_8, the G_{flavor} are,
respectively, $SU(3) \otimes SU(3)$, $SU(6)$, and $SU(3) \otimes SU(3) \otimes$
$SU(3)$. [Incidentally these maximal subgroups are the lit-
tle groups of $(1,e_7)$].

Let me illustrate the use of these groups in the case
of E_6: the smallest nontrivial representation is $\underset{\sim}{27}$, which
decomposes under $SU_L(3) \otimes SU_R(3) \otimes SU_c(3)$ as

$$(3,\bar{3},1^c) \oplus (3,1,3^c) \oplus (1,\bar{3},\bar{3}^c) :$$

the nine color-singlets are leptons, some of which may be
Majorana neutrals; the middle factor consists of three
flavors of left-handed colored quarks; the last, three
flavors of left-handed antiquarks (i.e., antiparticles of
the right-handed quarks). In order to implement the GIM

scheme, one has to postulate another 27. The Weinberg-
Salam $SU_W(2) \otimes U(1)$ is a subgroup of $SU_L(3) \otimes SU_R(3)$ where
the $SU_W(2)$ is embedded in $SU_L(3)$.

There are other candidates for G that have been discussed in the literature, for example, SU(5) of Georgi
and Glashow,[25] and SO(10) in connection with the vectorlike model.[26] It may or may not be relevant that SU(5) is
isomorphic to E_4 (the groups in the E series are all exceptional except E_4; E_4 fails to be exceptional because
of this isomorphism). Incidentally, all these groups are
anomaly-free.

XIV. STABILITY OF THE PROTON

The lifetime of the proton is at least 10^{30} years.
This is to be compared with the age of the universe, about
$10^{11\pm1}$ years. All supergroup schemes contains a germ for
instability of the proton, and let me explain this.

In the usual picture, the proton consists of three
quarks, completely antisymmetric in color. Therefore any
two of them are antisymmetric in color (these are the
diquarks): they transform like $\bar{3}^c$ and are bosonic. They
cannot go into an antiquark, but they can go into $\bar{q} + \ell$
(leptoquark) because the lepton ℓ is color-singlet, and
the leptoquark is bosonic:

Diquark	$\leftrightarrow$	Leptoquark
$(qq)_{antisym}$		$(\bar{q}\ell)$

Of course this group-theoretic argument is true whether
or not there is a supergroup G. It is just that in schemes
where leptons and quarks belong to separate multiplets,
there is no agent which can mediate this transition.

In super-unified schemes though, some of the gauge
bosons corresponding to the coset $G/[SU_c(3) \otimes G_{flavor}]$
carry the quantum numbers of diquarks/leptoquarks. Thus
we cannot avoid, unless something intervenes, there being
a process like $p \to \pi^+ + \nu$ which I picture as

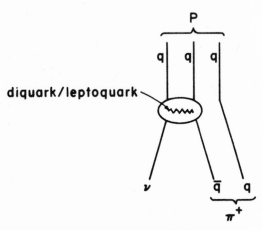

      Murray Gell-Mann suggested three "cures" for this.
We can arrange the dynamics so that this process is
"postponed" to higher orders in G_F, or we can make the
masses of these gauge bosons very massive, of the order
of the Planck mass. Lastly, define the baryon number N
as $N_{local} + N_{global}$, where N_{local} is associated with a
local gauge invariance, N_{global} with a global gauge
invariance; arrange the matters (presumably through vacuum
expectation values of the Higgsians) so that each is violated, but not the sum, which is the physical baryon
number.

      It is intriguing (and mind-boggling) to speculate on
what would happen to this issue if we had a respectable
octonion field theory. You see, the coset $G/[SU_c(3) \times$
$G_{flavor}]$ does not leave e_7 invariant, the e_7 which is the
analogue of the i in quantum mechanics, and the gauge
bosons associated with this coset may have a completely

different physical significance.

XV.   HIERACHICAL SYMMETRY BREAKING AND HIGGSIANS

How does the super symmetry G break down to the "observed" symmetry $SU_c(3) \otimes U_W(2)$?   According to a more conservative picture in which all spontaneous breakdowns are due to Higgs phenomena the scenario is something like the following;[27] we have to bear in mind in this discussion the theorem due to Tom Appelquist and Jim Carazzone,[28] which states that in studying a phenomenon on the mass scale M, effects of particles of masses $\gg$ M may be completely neglected.

Let M be the running mass scale in the sense of the renormalization group analysis.   In the regime $M \gtrsim 10^{19}$ GeV or the Planck mass, there is a manifest supersymmetry G, and the associated unified coupling constant g(M), which presumably becomes logarithmically weaker with increasing M.   In the intermediate range 10^3 GeV$<$M$\ll 10^{19}$ GeV the effects of diquark/leptoquarks may be neglected and the apparent symmetry is that of $SU_c(3) \otimes G_{weak}$.   Below $M \approx 10^3$ GeV, the presently observed pattern sets in, with a broken $U_W(2)$ describing weak and electromagnetic interactions, and an exact $SU_c(3)$ describing strong interactions.   The former is described by an infrared fixed-point theory, the latter by an ultraviolet fixed-point (asymptotically free) theory.   I have sketched in the following figure how various coupling strengths vary as the running mass scale varies.   Dotted portions in the figure are conjectural.   They depend on what the groups in question are, and how many fermions and relevant Higgs mesons there are at each mass scale.

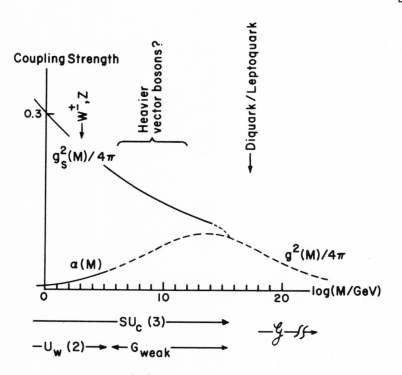

The Higgs meson(s) responsible for the spontaneous breakdown of $U_W(2)$, if it exists, is relevant to today's observational physics. The phenomenology of the Higgsian meson was recently surveyed by Ellis, Gaillard and Nanopoulos.[29] Steve Weinberg[30] has, more recently, shown that the mass of the Higgs meson is ≥ 4 GeV.

Dave Cline[31] argues that the best way of producing Higgs mesons (H°) may be in the ν-N scattering via the following mechanism:

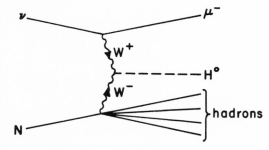

A naive estimate of the H° production ratio gives[32] about 10^{-5} in neutrino interactions.

While these topics were not dealt with in this conference, they form an integral part of my own perspective on particle physics.

XVI.   MAGNETIC MONOPOLE

The beautiful Maxwell equations can be improved upon in at least two ways. Professor Dirac told us how he came upon the magnetic monopole. He wanted a more symmetric system of equations, a system symmetric with respect to $E \leftrightarrow B$. The compatibility of this system with quantum mechanics, specifically the single-valuedness of wave functions, led to electric charge quantization.

The second way is to embed the local gauge invariance of the Maxwell theory in a compact non-abelian simple group, say SO(3), and devise a system of equations which is covariant under the local non-abelian gauge transformations. This second approach leads to the Yang-Mills theory. In this scheme electric charge quantization is automatic, since the topology of the U(1) associated with electric charge conservation is compact.

There is a strong connection between these two approaches. On the one hand, Polyakov and 't Hooft have shown that in a SO(3) gauge theory with scalar fields, there is a soliton-like solution which exhibits all the characteristics of a Dirac magnetic monopole. This monopole is very massive, $M \sim 10^4$ GeV.

On the other hand, Fred Goldhaber showed us another connection. If we follow Dirac's original line of thought, we see the correspondence between the quantized quantity eg and a component of the angular momentum, where g is the magnetic charge.

    Goldhaber proposed to interpret it as the third
component of the isospin, and to modify the Hamiltonian
(by adding more gauge fields and interactions amongst
them) so as to make it commute with all three components
of the isospin. The outcome is the Yang-Mills theory.
Fred gives credit to Henri Poincaré for this insight;
I am sure he is being very modest.

    I believe that magnetic monopoles do exist (but it
is just a belief, nothing more), perhaps at a very large
mass, because such a beautiful mathematical construct
must have a counterpart in nature. I would at this
juncture call your attention to a series of papers by
Wu and Yang[33] on the connection among gauge theories,
magnetic monopoles, and the branch of mathematics known
as fibre bundles. (The "bundle" here has nothing to
do with the same word used in Professor Dirac's lecture).

    XVII.   QUARK CONFINEMENT-PERTURBATIVE APPROACH
    The hope for quark confinement in QCD arises from
the following circumstance. The massless gluon theory
has a severe infrared disease; the effective coupling is
getting stronger as the mass scale M goes to zero. It may
be that the theory circumvents this difficulty by allowing
only colorless objects as asymptotically observable particles; in configuration space, two quarks cannot be separated by a macroscopic distance, because the force acting
between them grows progressively as the separation gets
bigger.

    There has always been a stumbling block in pursuing
this idea, and it is known as the Kinoshita-Lee-Nauenberg
theorem. Since I am more familiar with the work of T.D.
Lee and Nauenberg, let me cast the discussion in their
language. Suppose we have a Hamiltonian $H(\mu)$, which

exhibits a degenerate spectrum in the limit $\mu \to 0$. We consider an inclusive cross section $\sigma^{(n)}(\mu)$, computed in the n-th order in the coupling constant. The inclusiveness of the cross section entails summing over ensembles of initial and final states which would become degenerate if μ were zero. The theorem states that

$$\lim_{\mu \to 0} \sigma^{(n)}(\mu) = \text{finite} \quad .$$

Superficially this seems to imply that it is not possible to find signs of quark confinement in perturbation theory.

Tom Appelquist reported on a long calculation performed by him, Jim Carazzone, Hannah Kluberg-Stern and Mike Roth, the latter three at Fermilab, to lowest non-trivial order, to verify this theorem. A related circulation was also performed by Ed Yao at Michigan. Both groups confirm the Kinoshita-Lee-Nauenberg theorem in a non-abelian gauge theory. In particular, the former group shows that with a detector of finite energy resolution which does not see color, it is possible, for example, to detect a quark by its fractional charge--this if perturbation theory makes sense.

On the other hand, Mike Cornwall reported on his rather extensive and painstaking work with George Tiktopoulos, which indicates that color confinement does occur in QCD, and the S-matrix elements involving external colored objects are strongly damped to zero as $\mu \to 0$. I cannot vouch for the technical correctness of their calculation. I can say, however, that there is no contradiction between Cornwall's and Appelquist's discussions. The UCLA group arranges the calculation so that they sum over leading log μ terms to all orders in perturbation

theory first, and then let the parameter μ vanish: in
other words, they are calculating in principle the following:

$$\sigma = \lim_{\mu \to 0} \sigma(\mu) = \lim_{\mu \to 0} \left[\sum_{n}^{\infty} \sigma^{(n)}(\mu) \right] \quad .$$

The inference that I draw from these discussions is that
the summation and the limiting process μ→0 do not commute.

I shall argue that the Cornwall-Tiktopoulos strategy
is a plausible one if color confinement is anything like
a phase transition in many body problems, in the next
section.   In the meantime, please note that the finite μ
serves two purposes.   They put in μ^2 in the Feynman-gauge
gluon propagator by hand;

$$g_{\mu\nu} \frac{1}{k^2 + i\varepsilon} \to g_{\mu\nu} \frac{1}{k^2 - \mu^2 + i\varepsilon} \quad .$$

It is an infrared cutoff, and it is also a device to break
gauge invariance explicitly.   I will make use of this observation in the next section.

XVIII.   QUARK CONFINEMENT AS A CRITICAL PHENOMENON

Ken Wilson described the progress he has made in the
lattice gauge theory.   In previous versions of the theory,
he succeeded in establishing color confinement as long as
the lattice constant a was kept finite.   In the new formulation which resembles the block spin method in statistical
mechanics, the S-matrix is in principle a-independent and
covariant.   We all await eagerly his final numerical results - meson and baryon masses, cross-sections, decay
rates, etc.

In the last month or two, I have been very fortunate to have my two esteemed friends -- Jean Zinn-Justin and Bill Bardeen -- explain to me their views on quark confinement as a critical phenomenon, and interpret for me the views of Wilson, Kogut and Susskind, and Migdal[34] and Polyakov.[35] I would like to describe to you my understanding of the subject; for my misunderstanding or misrepresentations I alone am responsible.

To cirvumvent the infrared catastrophe, let us work in $4 + \varepsilon$ dimensions, $\varepsilon > 0$. The β function of the renormalization group looks like the following figure:

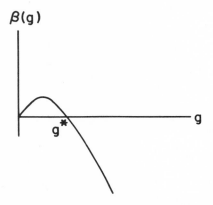

The ultraviolet fixed point $(g^*)^2$ is of order ε. For $g < g^*$, the perturbation expansion probably makes sense at least as an asymptotic expansion. Note that the gauge Lagrangian $L(A_\mu, g)$ can be written as

$$L(A_\mu, g) = \frac{1}{g^2} L(gA_\mu, 1)$$

and the analogy between g^2 and kT in statistical mechanics is complete: g^{*2} corresponds to the critical temperature T_c. Beyond g^*, the theory is in a different phase, and perturbation theory is no help at all here.

There is a way of overcoming this impasse. For

example, in a ferromagnetic system it is only necessary
to turn on an external magnetic field (in which case there
is no critical point in correlation functions) in order
that a perturbation series make sense. Another example,
perhaps more familar to particle physicists, is the σ
model:

$$L_\sigma = \tfrac{1}{2}\left((\partial_\mu \sigma)^2 + (\partial_\mu \pi)^2\right) - \frac{\mu^2}{2}(\sigma^2 + \pi^2) - \frac{\lambda}{4}(\sigma^2 + \pi^2)^2 \quad .$$

As long as $\mu^2(\sim T - T_c)$ is positive the expansion in λ
makes sense. To go to the other phase (the Goldstone
phase), one first adds to the Lagrangian a symmetry-break-
ing term, $c\sigma$ (the analogue of external magnetic field),
next develops a new perturbation series in λ(with $\lambda<\sigma>^2$
fixed), then continues μ^2 to a negative value, and finally
lets c go to zero.[36] One obtains in this way Green's
functions valid in the Goldstone phase. These procedures
are based on an assumed theorem (which has been proved in
some simple cases, and is known as the Simon-Griffiths
theorem) that as long as there is an explicit symmetry
breaking, Green's functions or correlation functions are
analytic (or at least asymptotic) in the dominant couplings
when subsidiary parameters are suitably chosen.

The procedure Cornwall and Tiktopoulos adopted, it
seems to me, is a variation of this general strategy.
They break the gauge symmetry explicitly, resume the perturbation series and then, and only then, remove the explicit symmetry breaking, to reach the "high temperature"
phase.

Zinn-Justin and Edouard Brézin[37] have studied, as
have Migdal and Polyakov, as I understand, the nonlinear
σ-model, in 2 + ε dimensions. It bears a striking resemblance to the gauge theory in many ways. One is that

at $\varepsilon = 0$, the nonlinear σ-model, in which $\sigma + i\underset{\sim}{\tau}\cdot\underset{\sim}{\pi} = f_\pi \exp(i\underset{\sim}{\tau}\cdot\underset{\sim}{\phi}/f_\pi)$, and which may be considered as a description of the Goldstone phase, has a bad infrared disease; the second is that the β-function has the same behavior as in the gauge theory. They have succeeded, by the aforementioned strategy, in reaching the normal phase in which $\langle\sigma\rangle = 0$.

Actually, the resemblance between the nonlinear σ-model and the gauge theory is more than skin-deep. The lattice gauge theory is invariant under $\prod_n \left[SU_c(3)\right]_n$, where n labels lattice sites: the gauge linkage $U_{n,\hat{\mu}} = \exp\left[iga \times \lambda\cdot\underset{\sim}{A}_\mu(n)\right]$ in Ken Wilson's lattice theory[38] is a nonlinear realization of $\left[SU_c(3)\right]_n \otimes \left[SU_c(3)\right]_{n+a\hat{\mu}}$, just as $f_\pi \times \exp(i\underset{\sim}{\tau}\cdot\underset{\sim}{\phi}/f_\pi)$ is a nonlinear realization of $\left[SU(2)\right]_L \otimes \times \left[SU(2)\right]_R$. Bill Bardeen is investigating the dynamical circumstances in which $U_{n,\hat{\mu}}$ takes a form other than the one written down above.

As $\varepsilon \rightarrow 0_+$, we expect to recover the real physical situation. From the work of Brézin and Zinn-Justin, it seems clear to me that the σ-model has only the normal phase in 2 dimensions, thus obviating the Coleman theorem which proscribes Goldstone bosons in 2 dimensions. By an obvious extension, I suspect that an asymptotically free gauge theory has only phases in which colored objects are confined.

There are several immediate questions on which we debate hotly. Firstly, what is the order parameter in QCD, which heralds the onset of a new phase? How does a bag picture of hadrons arise in a confinement phase? What happens to the super gauge theory based on G on a lattice, and are there other ways of breaking G down to $SU_c(3) \otimes G_{weak}$ in this setting? There are myriads of questions we can raise, or haven't thought about.

XIX. EPILOGUE

There are questions I haven't even begun to summarize, but I am using up very fast the allotted time. I feel a Voltaire tapping on my shoulder and whispering to me "Cela est bien dit, mais il faut cultiver notre jardin."[39] In contemporary American, I think it means "Shut your trap and go back to work."

We are living in an exciting era. Let's all get back to work. Thank you.

I thank Robert Shrock for his help in the preparation of the manuscript.

FOOTNOTES

These footnotes are personal annotations to the
text. They are not intended to be, nor should they be
used as, a bibliography. Please refer to individual
talks for authoritative references.

1. See, W. Pauli, Collected Scientific Papers, edited
    by R. Kronig and V. Weisskopf, (Interscience Publish-
    ers, New York, 1964), pp xiii xvi.

2. R. P. Feynman, in Neutrino-1974, edited by C. Baltay
    (American Institute of Physics, New York, 1974),
    pp 299 - 319.

3. See P. Nath's and B. Zumino's contributions to the
    Proceedings of the Conference on Gauge Theories and
    Modern Field Theory at Northeastern University,
    September, 1975 (to be published by the MIT Press).

4. For public pronouncement, C. N. Yang. "Some Concepts
    in Current Elementary Particle Physics," in The
    Physicist's Conception of Nature, edited by J. Mehra
    (D. Reidel Publishing Co., Dordrecht, Holland, 1973),
    pp 447 - 453.

5. See the summary of Leon Lederman in the forthcoming
    Proceedings of the Photon/Lepton Symposium at SLAC,
    1975.

6. E. G. Cazzoli, et al., Phys. Rev. Lett., $\underline{34}$, 1125
    (1975).

7. H. Deden, et al., Phys. Lett., $\underline{58B}$, 361 (1975); J.
    von Krogh, at the Irvine Conference, December 5 (1975);
    G. Blietszhau, et al., TCL/Int. 758, CERN preprint;
    J. von Krogh, et al., Wisconsin preprint. I wish to
    thank J. von Krogh and members of the Wisconsin-
    Berkeley-CERN-Hawaii collaboration for making avail-
    able to me their data prior to publication.

8. Y. Watanabe, et al., Phys. Rev. Lett., $\underline{35}$, 898 (1975);

L. Hand informed me that, to about $\pm 7 \sim 8\%$ absolute accuracy, the Berkeley-Cornell-Michigan State group data fit the empirical formula

$$F_2(\omega,Q^2) = F_2^{SLAC}\ (\omega^-)\left(\frac{\omega}{\omega_0}\right)^{b\left[\ln\ Q^2/(3\ GeV^2)\right]}\ ,$$

where

$$b = .090 \pm .018$$
$$\omega_0 = 6.1(+3.9,-2,4),\quad (statistical\ errors)$$

for $Q^2 \geq 3\ GeV^2$. The best fit is obtained if the Stein fit in ω^- is used for F_2^{SLAC} (S. Stein, et al., Phys. Rev. D12, 1884 (1975). I wish to thank Lou Hand for presenting the data in a form that I can digest.

9.   I. Karliner, "The effect of Heavy Lepton on Charmed Particles Decays", IAS preprint (1975).

10. I. Karliner and B. W. Lee, private discussion.

11. See in this connection, J. Kandaswam, J. Schecter and M. Singer, "Possible Enhancement of the Leptonic Decays of Charmed Pseudoscalars," Syracuse preprint (1975).

12. See Peter Minkowski's talk in these Proceedings, and the discussion in a later section.

13. See for example M. K. Gaillard, B. W. Lee and J. L. Rosner, Revs. Mod. Phys. 47, 277 (1975).

14. E. Golowich and B. Holstein, Phys. Rev. Lett. 35, 831 (1975).

15. Pakvasa and H. Sugawara, Hawaii preprint (1975).

16. L. Maiani, "CP Violation in Purely Lefthand Weak Interactions," Rome preprint (1975).

17. M. K. Gaillard and B. W. Lee, Phys. Rev. Lett., 33, 108 (1974). G. Altarelli and L. Maiani, Phys. Lett., 52B, 351 (1974).

18. My recent readings on this subject include: G. Farrar, Phys. Rev. D4, 212 (1971); M. K. Gaillard, Nuovo Cimento 6A, 559 (1971). For a more complete bibliography see the papers A. Zee cited.

19. This is a tribute to M. K. Gaillard. I thank her for intense discussions on this point during the SLAC Photon/Lepton Symposium in 1975. See also footnote 20 below, especially the note added in proof.

20. J. Ellis, M. K. Gaillard, and D. V. Nanopoulos, Nuclear Physics B100, 313 (1975).

21. I believe the following points were first made in the literature by M. K. Gaillard, "Charm," in The Proceedings of the Xth Rencontre de Moriond, edited by J. Tran Thanh Van, (CRNS publication, 1975); see also A. Pais and S. B. Treiman, Phys. Rev. Lett., 35, 1556 (1975); talks of L. Wolfenstein and C. Llewellyn-Smith at the SLAC Photon/Lepton Symposium, 1975.

22. The characteristic x and y distributions of the dimuon events in the charm scheme were emphasized in B. W. Lee, "Dimuon Events," in the forthcoming Proceedings of the Conference on Gauge Theory and Modern Field Theory (to be published by the MIT Press).

23. See the talk of B. Roe in the forthcoming Proceedings of the Photon/Lepton Symposium at SLAC, 1975.

24. F. Gürsey, "Color Quarks and Octonions," in The Johns Hopkins University Workshop on Current Problems in High Energy Particle Theory, 1974. G. Domokos and S. Kövesi-Domokos, (eds.) pp 15-42 (Physics Department, The Johns Hopkins University).

25. H. Georgi and S. L. Glashow, Phys. Rev. Lett., 32, 438

(1974).

26. H. Fritzsch, M. Gell-Mann and P. Minkowski, Phys. Lett., <u>59B</u>, 256 (1975).

27. H. Georgi, H. R. Quinn and S. Weinberg, Phys. Rev. Lett., <u>33</u>, 451 (1974).

28. T. Appelquist and J. Carazzone, Phys. Rev. <u>11</u>, 2856 (1975).

29. J. Ellis, M. K. Gaillard and D. V. Nanopoulos, "A Phenomenological Profile of the Higgs Boson," CERN preprint (1975).

30. S. Weinberg, "Mass of the Higgs Boson," Harvard preprint.

31. D. Cline, in a luncheon conversation.

32. M. K. Gaillard, private communication.

33. T. T. Wu and C. N. Yang, "Concept of Nonintegrable Phase Factors and Global Formulation of Gauge Fields" and "Some Remarks about Unquantized Non-Abelian Gauge Fields," to be published; C. N. Yang, Phys. Rev. Lett., <u>33</u>, 445 (1974).

34. A. A. Migdal, "Gauge transitions in Gauge and Spin Lattice Systems," L. D. Landau Institute for Theoretical Physics preprint (1974).

35. A. M. Polyakov, Phys. Lett. <u>59B</u>, 79 (1975).

36. See for example, B. W. Lee "Chiral Dynamics," (Gordon and Breach, New York, 1972).

37. E. Brézin and J. Zinn-Justin, "Renormalication of the Nonlinear σ-model in 2 + ε Dimension -- Application to Heisenberg Ferromagnets," a private note.

38. The best reference may be K. Wilson's lectures in the forthcoming Proceedings of the Erice Summer School, 1975.

39. Voltaire, in <u>Candide</u>, <u>ou</u> <u>l'Optimisme</u>.

PARTICIPANTS

Carl H. Albright
Northern Illinois University

Thomas Appelquist
Yale University

Marshall Baker
University of Washington

William Bardeen
University of Wisconsin

Vernon Barger
University of Wisconsin

Isac Bars
Yale University

P. N. Bogolubov, P.N.
Ins. for Nuclear Research
Academy of Science U.S.S.R.

Richard Brandt
New York University

Laurie Brown
Northwestern University

Arthur Broyles
University of Florida

Nina Byers
University of California

Robert Cahn
University of Washington

Peter Carruthers
Los Alamos Scientific Lab.

George Chapline
Lawrence Livermore Lab.
University of California

M. Chen
Massachusetts Institute of
  Technology

Fred Cooper
Los Alamos Scientific Lab.

John Cornwall
University of California--
  Los Angeles

Michael Creutz
Brookhaven National Lab.

Richard Dalitz
Oxford University

Stanley Deans
University of South Fla.

Alvaro de Rújula
Harvard University

P. A. M. Dirac
Florida State University

Max Dresden
State University of New
  York--Stony Brook

Loyal Durand
Institute for Advanced
  Study--Princeton

Glennys Farrar
California Institute of
  Technology

Gordon Feldman
John Hopkins University

Paul Fishbane
University of Virginia

Peter Freund
University of Chicago

Harald Fritzsch
California Institute of
  Technology

399

Frederick Gilman
Stanford University

Sheldon Glashow
Harvard University

Alfred Goldhaber
State University
 of New York Stonybrook

Ahmad Golestaneh
Fermi National Accelerator
 Laboratory

Christian Le Monnier
 de Gouville
Center for Theoretical
 Studies
University of Miami

O. W. Greenberg
University of Maryland

Feza Gürsey
Yale University

Alan Guth
Columbia University

C. R. Hagen
University of Rochester

Leopold Halpern
Florida State University

M. Y. Han
Duke University

Joseph Hubbard
Center for Theoretical
 Studies
University of Miami

Muhammad Islam
University of Connecticut

Ken Johnson
Massachusetts Institute of
 Technology

Lorella Jones
University of Illinois

Gabriel Karl
University of Guelph

T. E. Kalogeropoulos
Syracuse University

Abraham Klein
University of Pennsylvania

Behram Kursunoglu
Center for Theoretical
 Studies
University of Miami

Willis Lamb
University of Arizona

Benjamin Lee
Fermi National Accelerator
 Laboratory

Don Lichtenberg
Indiana University

K. T. Mahanthappa
University of Colorado

Alfred Mann
University of Pennsylvania

André Martin
CERN

Caren Ter Martirosiyan
Institute for Theoretical
 and Experimental Physics
Moscow, U.S.S.R.

V. A. Matveev
JINR, Moscow, U.S.S.R.
 and Fermi Laboratory

Meinhard Mayer
University of California-
 Irvine

Barry McCoy
SUNY, Stonybrook

Sydney Meshkov
National Bureau of Standards

Peter Minkowski
California Institute of
  Technology

J. W. Moffat
University of Toronto

Yoichiro Nambu
University of Chicago

Yuval Ne'eman
Tel-Aviv University

André Neveu
Institute for Advanced
Studies
Princeton

Roger Newton
Indiana University

Kazuhiko Nishijima
University of Chicago

Richard Norton
University of California -
  Los Angeles

Horst Oberlack
Max Planck Institut for
  Physics and Astronomy
Munich, West Germany

Reinhard Oehme
University of Chicago

Lars Onsager
Center for Theoretical
  Studies
University of Miami

Heinz Pagels
Rockefeller University

Sandip Pakvasa
University of Hawaii

Michael Parkinson
University of Florida

Jogesh C. Pati
University of Maryland

R. D. Peccei
Stanford University

Arnold Perlmutter
Center for Theoretical
  Studies
University of Miami

H. David Politzer
Harvard University

P. G. Price
University of California -
  Berkeley

Pierre Ramond
California Institute of
  Technology

Rudolf Rodenberg
III Physikalisches Institut
  der Technischen Hochschule
  Aachen, West Germany

Fritz Rohrlich
Syracuse University

S. P. Rosen
ERDA

Ronald Ross
University of California -
Berkeley

V. I. Savrin
Institute for High Energy
  Physics
Serpukhov, U.S.S.R.

Howard Schnitzer
Brandeis University

Frank Sciulli
California Institute of
 Technology

Gordon Shaw
University of California
 Irvine

Dennis Silverman
University of California -
 Irvine

Alberto Sirlin
New York University

L. Slavnov
Institute of Math. of the
 Academy of Sciences
Moscow, U.S.S.R.

L. Soloviev
Serpukhov Institute for
 Theoretical Studies
Serpukhov, U.S.S.R.

George Soukup
Center for Theoretical
 Studies
University of Miami

Joseph Sucher
University of Maryland

Katsumi Tanaka
Ohio State University

William Tanenbaum
Stanford Linear Accelerator
 Center

John R. Taylor
University of Colorado

Vigdor Teplitz
Va. Polytechnic Institute
and State University

George Tiktopoulos
University of California -
 Los Angeles

Yukio Tomozawa
University of Michigan

T. L. Trueman
Brookhaven National Lab.

V. S. Vladimirov
Institute of Math. of the
 Academy of Sciences
Moscow, U.S.S.R.

Kameshwar C. Wali
Syracuse University

Geoffrey West
Los Alamos Scientific Lab.

Ken Wilson
Cornell University

Lincoln Wolfenstein
Carnegie-Mellon University

T. T. Wu
Harvard University

G. B. Yodh
University of Maryland

Fredrick Zachariasen
California Institute of
 Technology

Anthony Zee
Princeton University

Daniel Zwanziger
New York University

SUBJECT INDEX

Date Due

			UML 735